MAKING THE DIGITAL CITY

Making the Digital City
The Early Shaping of Urban Internet Space

ALESSANDRO AURIGI
University of Newcastle upon Tyne, UK

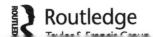

Routledge
Taylor & Francis Group

First published 2005 by Ashgate Publishing

Published 2016 by Routledge
2 Park Square, Milton Park, Abingdon, Oxon OX14 4RN
711 Third Avenue, New York, NY 10017, USA

Routledge is an imprint of the Taylor & Francis Group, an informa business

British Library Cataloguing in Publication Data
Aurigi, Alessandro
  Making the digital city : the early shaping of urban
  Internet space. - (Design and the built environment series)
  1.City planning - Technological innovations - European
  Union countries 2.Information technology - Social aspects -
  European Union countries 3.Cities and towns - European
  Union countries - Information services 4.Metropolitan areas
  - European Union countries - Information services
  5.Internet - Social aspects - European Union countries
  6.Cyberspace - European Union countries 7.Public spaces -
  European Union countries
  I.Title
  307.1'216'094

Library of Congress Cataloging-in-Publication Data
Aurigi, Alessandro.
 Making the digital city : the early shaping of urban internet space / by Alessandro
Aurigi.
     p. cm. -- (Design and the built environment)
 Includes bibliographical references and index.
 ISBN 0-7546-4364-6
 1. City planning--Technological innovations--European Union countries. 2.
Information technology--Social aspects--European Union countries. 3. Cities and
towns--European Union countries--Information services. 4. Metropolitan areas--
European Union countries--Information services. 5. Internet--Social aspects--European
Union countries. 6. Cyberspace--European Union countries. 7. Public spaces--European
Union countries. I. Title: Digital city. II.
Title: Urban Internet space. III. Title. IV. Series: Design and the built environment
series.

 HT169.E8A95 2005
 307.1'216'02854678--dc22   2005009756

ISBN 9780754643647 (hbk)
ISBN 9781138276536 (pbk)

Contents

List of Figures

List of Tables

Preface

Urban technology is a field which, comparatively with other threads of urban studies, is relatively new and in need of exploration. But it is also a powerful generator of meanings, symbols and visions, and in the past couple of decades the gradual emergence and diffusion of Information and Communication Technologies in our everyday lives has catalysed the promotion of a variety of views of urban futures, and a plethora of buzzwords linked to these. Different conceptions of the articulation between cities and telematics, and of aspects of this, throughout recent history have been characterised by their own particular definitions. Graham and Marvin give examples of these in their 1996 book *Telecommunications and the City*, gathering a list ranging from 'the informational city' to 'teletopia'. Between the end of the 1980s and the beginning of the 1990s urban ICT projects were usually labelled with the prefix 'tele', such as the British and French 'teleports', and tended to be characterised by being still envisaged as centralised facilities connecting mainframe computers and media centres in different cities. In the second half of the 1990s, though, the popularisation of the Internet, previously existing only as a restricted academic tool, introduced new visions and possibilities for the city. The Internet was by definition a non-centralised or controllable environment, it was the sum of its individual, smallest parts or, better, more than it. Together with it came the coining of the term – and the idea – of 'cyberspace' and the 'virtual community'. Teleports were symbolically out of fashion, and the hype broke out on the potential of the more complex and somehow anarchic movement of civic networking, the American 'FreeNets' and in general virtual – or 'digital' – public space. Soon 'proper' cities, local authorities and often partnerships of public and private entities got interested in the potential of the net for empowering citizens. They plunged into the world of what in Europe has been known as the 'digital city', a more or less complex urban information and communication system, often supported by the presence of a web portal.

This is a critical history of the early steps of the digital city, an attempt to look at the phenomenon by concentrating on aspects that go beyond the huge amount of hope – and hype – that was generated in those years about the potential of the net for revitalising the urban public sphere.

The book achieves – in my very humble opinion – two purposes. The first and most obvious is providing a snapshot on the early history of the proactive, somehow planned, deployment of what could be seen as digital public space in European cities. It explores what was done and achieved by trying to go well beyond the glossy hi-tech, future-bound images that ever-competing cities would try and project to the outside world.

The second is to prove that although urban ICT can develop fast in terms of more advanced technologies, bandwidth availability, and programming codes, this has to come to terms with the way these systems and 'solutions' are socially shaped, and are heavily affected by the urban reality in which they develop. It is a rather common myth thinking that high technologies grow and change almost too fast to be studied properly, and when a book about technology is released, it will inevitably be already seriously out of date. This can be a valid assumption when considering works that deal with the details of the short-lived 'latest' release of a certain piece of software or technical innovation. However, when it comes to considering the processes by which innovative technologies are deployed into, and adapted to, the urban environment of our cities, matters become incredibly more complex and slow-moving, and if technology is supposed to affect – and change – the city, it is also very true that the city affects technology.

This book looks above all at these processes, and the interpretations that have been leading them, and because of this perspective can enjoy a definitely longer 'shelf-life' than one could reasonably expect from an essay on civic websites. Indeed, the study described here will try and show to the reader that principles and dilemmas which were valid nearly ten years ago, for the early civic portals that were being launched in the second half of the 1990s, are still very relevant now, for the 'digital' city of the 21st century and its more advanced and versatile ICT-based ideas, projects and spaces.

Alessandro Aurigi
Newcastle upon Tyne, 2005

Acknowledgements

The research behind this book would have never been possible without the help and encouragement of far too many people. Among the academic colleagues, I wish to thank in particular Stephen Graham from Durham University, Allan Gillard, the former Head of Department of Town and Country Planning at Newcastle, and Paolo Scattoni from La Sapienza University in Rome.

I owe a great deal to the kindness and helpfulness of my 'gatekeepers' in Bologna and Bristol, the project managers and senior researchers without whom it would have been impossible to access information, crucial opinions, and ultimately understand better the digital city phenomenon.

Last but not least a big thank you to my wife, daughter and parents for bearing with me so well.

About the Author

Alessandro (aka Alex) Aurigi holds a Laurea in Architecture from Florence University (Italy) and a PhD from Newcastle University (UK). He currently works as a lecturer at Newcastle University, where he is a member of the Architectural Informatics (AI) research group, and an associate member of the Global Urban Research Unit (GURU).

Previously, he has worked as a lecturer at the Bartlett School of the Built Environment, University College London, and as a research fellow in the Centre for Advanced Spatial Analysis (CASA), UCL. He was also a member of the Centre for Urban Technology (CUT) at Newcastle University

His main research interest is studying the relationships between the emergence of the 'information society' and the ways we imagine, conceive, represent, and manage buildings and cities.

PART I
THE CONTEXT:
PUBLIC SPACE AND CYBERSPACE

Chapter 1

Introduction

Introduction

This book is a critical history of the emergence of 'civic' Internet information systems in Europe, their role as innovative regeneration 'tools', and above all how they have been socially constructed. This history starts from the mid of the 1990s.

The 1990s, and in particular their second half, were characterised by a surge in technological advances in Information and Communications Technologies (ICTs). Among these were increasing data transmission speeds, rapid evolution of microprocessors and their computing power, and possibly above all the emergence of the World Wide Web, an accessible way of distributing and utilising multimedia information over the Internet. It soon appeared to the IT industry as well as a range of commentators that the web definitely had the potential of popularising the Internet, making its focus and mission shift from that of a mainly academic international networking system, to a multi-purpose – and nearly all-purpose – global information grid.

This is exactly what started happening, as the number of Internet hosts – the computers permanently connected to it – started growing exponentially, together with the amount of information available on the new web platform.

Such a phenomenon was rather obviously bound to be seen as a 'revolution' of analogous importance to other previous milestones in human social history. Books started appearing arguing that, after the industrial revolution, humankind was facing an information revolution of vast proportions and significance, from business related texts such as Frances Cairncross' *The Death of Distance* (1998), to writings in Architecture History such as Charles Jencks' *What is Post-Modernism?* (1986 and revised in 1996), with its section on the 'Post-Modern Information World'. The conviction that society had to be interpreted as post-fordist, post-modern, information and knowledge-based, was being reinforced more than ever.

Society was seen by many commentators to be changing fast and radically, partly because of the heavy impacts of the 'layer' of electronic information and its potential for 'virtualising' functions that before would have been affected and shaped by the traditional limitations of physical distance, as well as time.

Hype was dominant, as the enthusiasm for the potentially beneficial impacts of new technologies on all aspects of society was contagious and hi-tech and telecommunications companies were for obvious reasons proactively encouraging this attitude.

Visions of a future technological 'heaven' were juxtaposed to fears of a dystopian perspective on a socially polarised society, more unjust than ever, heavily controlled and commodified.

Cities would have a central position in this wider debate, as ICTs were seen as revolutionising both spatial and socio-economic relationships, the main ingredients of urban environments. Again, even here, hype and speculation were prevailing, and even scholars and authoritative commentators were easily being carried away with making either enthusiastic or catastrophic predictions about urban futures where high technology was dominant.

Within this climate, newspapers and the media in general had started using terms like 'cybercity' or 'virtual city' to identify all sorts of – often very diverse – early experiments and ideas involving ICTs and cities. Martin Jacques from *The Guardian* newspaper visited Kuala Lumpur in 1997 and argued 'Modern planning is not just about roads and estates. It's about an "intelligent network" linking our offices and homes' (Jacques, 1997).

In the same year, *The Independent* was reporting on a forthcoming experiment – sponsored by the two IT giants Gateway and Microsoft – to be held in a secret street in Islington, London, where the whole neighbourhood was going to be wired and provided with computers and software to communicate, becoming an 'Internet Street' (Arthur, 1997).

In their timely book *Telecommunications and the City*, Stephen Graham and Simon Marvin highlighted the many different metaphors – and in a way buzzwords – used by scholars and commentators to describe the increasingly telecommunications-based urban space of western cities (Graham and Marvin, 1996, p.72).

However, a concrete possibility of using the Internet to help regenerating urban environments seemed to exist. Several early and strongly symbolic examples, often referred to as 'digital cities', 'virtual cities', or 'civic networks' were opening the way of what seemed to be a new 'movement' involving the implementation of civic cyberspace to benefit the physical city and its socio-economic environment.

These new urban projects could be seen as potentially beneficial towards economic development and regeneration, by providing innovative tools for place marketing on the one hand, and an environment facilitating the growth of a better informed, up-to-date and clued up workforce on the other. Other benefits could be envisaged within the complex field of city management, as the Internet could help delivering services – old and new – supporting transactions, conveying information to the population. Another extremely important area was the support of participation and public discourse, which would enhance the public sphere of fragmented Western cities and boost democratic processes.

Urban cyberspace could then be seen as a crucial piece of technological innovation which could be able to deliver benefits in all those areas, and in a way become a sort of parallel, even enhanced public space of Western cities, complementing and revitalising the physical public space, seen by many as a feature in deep crisis.

In the 1990s urban Internet was a novelty, and it attracted considerable amounts of funding from the European Community, making possible the development of some apparently very complex projects. Current, successful paradigms seemed to point towards a model of 'holistic' intervention, involving efforts of creating general, all-encompassing web sites which would present themselves as informational 'mirrors' of their host cities, and would therefore tend to address a variety of urban regeneration issues and problems, under the same 'digital city' label.

Within this scenario, the 'digital city' concept – though still used sometimes to label virtually any kind of urban-related IT project – was quickly associated mainly with a specific type of initiative: web-based urban information systems and virtual communities. This was probably partly due to the success of the most famous 'digital city' paradigm in the whole of Europe: the Amsterdam-based DDS, De Digitale Stad.

In a short amount of time many other 'digital cities' – or 'civic networks' as some would call them – were born all over Europe, constituting a potentially very interesting phenomenon of innovation in city management, economic regeneration, and community building. They could be seen as an attempt to establish a parallel public space – a cyberspace – in Western, fragmented urban societies. But these initiatives were very scarcely known or analysed, and their scope, characteristics, impacts, motivations, were generally taken for granted within the hype-ridden climate of the 1990s.

Little empirical research was accompanying these developments, and a clear need was felt to gain more precise knowledge of the phenomenon, an awareness which could go well beyond the hype present in the media and boosted by the interests of the numerous growing new technology businesses and firms.

The 'digital city' was here, but what was it really? What types of initiatives could be identified, and how was urban cyberspace being shaped? Objectives of a research on this topic had to be gaining an understanding of the scope and quality of the phenomenon, at the wider European scale, but also exploring the ways some 'best practice' examples were actually being constructed and managed. Were digital cities really 'holistic' or were they focused on some precise purpose? What actual potential for regeneration did these projects have, and were they being shaped to realise that potential? What was the vision behind some of them, and was it really compatible with the 'new public space' vocation that most commentators would assign to digital cities?

This book is based on empirical research done to explore the early stages – which tend to be also the most crucial ones – of development of web-based urban information systems – or digital cities – in Europe. These projects started appearing in the mid 1990s and provoked quickly a series of press reports and – in general – a great deal of hype on their potential. However, the most resilient – and possibly inspired – of them have in different ways survived the turn of the century, and have become an accepted, innovative aspect of urban life in their own cities. So, the book looks at the development of digital cities in the European Union at different stages of their ongoing evolution, embracing events happened between 1996 and 2004.

The narrative starts from an attempt at defining the phenomenon as a whole and recognising a series of major issues. It then concentrates on creating an early typology of digital cities, and then moves to the analysis of two exemplar case studies. As one of these projects has become an award-winning paradigm of a successful and complex 'digital city', its history has been updated to 2004. It is a story of civic web sites, and their configuration. It is an analysis of their features and contents, as well as a reflection on the processes and actors that were and are shaping them, and what this could mean for the realisation of their alleged beneficial impact on the planning of Western cities.

What the book is about

The main focus of this book is to gain an insight on the phenomenon of digital cities. The research it is based on has had to deal with a variety of ways employed by civic institutions – both public and private – to try and exploit the World Wide Web of the Internet to benefit specific aspects of urban regeneration, as well as providing 'platforms' where the growing virtual fragments belonging to cities could be organised, presented, and given sense.

On the one hand this has involved exploring uncharted territory, something about which very little had been written or documented properly. This has meant searching for basic knowledge and awareness on the topic, stemming from literature that was dealing with the wider issues of ICT and society, in order to produce an early – but strongly needed – theoretical framework for the interpretation of the phenomenon. What really were digital cities? Could they be considered, as the hype from the press seemed to take for granted, as an identifiable new trend of urban development? How many of these virtual sites could be found in Europe, and what were their characteristics? What were digital cities offering to the inhabitants of their physical counterparts? To what extent were crucial issues of inclusiveness and social polarisation in the access and use of technology considered and addressed by these initiatives?

On the other hand, as will be clarified in the first two chapters, concentrating on the immediately observable contents of some initiatives, however indispensable and desirable, has not been perceived as a complete enough method of studying the phenomenon. If nothing else, as technological development based on the Internet seemed destined to keep changing very quickly, some 'deep', qualitative approach to studying this topic, going beyond the contents in order to observe some of the underlying social factors affecting digital city design and management, seemed like a very useful and enriching approach.

Valuable and stimulating input from the sociology of technology, and its theories on the social shaping of technology, have added a very valuable dimension to this book.

How the book is organised

This book is organised in sections, which correspond to the phases of the investigative journey it outlines.

Chapters 2 and 3 deal with issues on cities, the urban public sphere, and recent technological developments in the domain of Information and Communication Technologies. Between the mid 1990s and the early 2000s, not much has been written on the specific topic of digital cities and urban planning. So, it has been quite natural to consider a wider perspective, and engage in wider literature about cities, telecommunications and the urban public sphere, in order to get suggestions and inspiration.

The fourth chapter deals with the initial exploration of the phenomenon, and provides results and answers relative to the need to identify the scope of the digital city 'movement' and how such initiatives could be categorised and analysed from the point of view of their contents, throughout the European Union. This is achieved through a survey-based approach, centred on the observation of many 'web cities' and their contents.

Chapters 5 and 6 describe the qualitative investigations of two exemplar, leading edge, digital cities. These are linked to and extend from the survey presented in chapter 3, which provides for the choice, from a spectrum of over 200 possible cases, some examples that are – or have been – likely to be considered as 'best practice' ones. The Bologna digital city in Italy – named 'Iperbole' and the Bristol one both complied with the survey's criteria for advanced examples, and scored high within it.

Chapter 7 provides a comparison of the two examples, extrapolating common trends and differences, and identifying a series of main issues, problems, and possible solutions that are related to the processes underlying the shaping of urban high technology initiatives, and can inform similar projects.

Chapter 8 wraps everything up into a series of conclusive remarks, highlighting some important changes in the way advanced telecommunication and Internet technologies are perceived, that are happening after the turn of the millennium. These changes confirm, on the one hand, the relevance and timeliness of works such as the investigations presented in this book, and on the other hand suggest the need for further, more refined research on the topic of urban telematics.

Chapter 2

Cyberspace and the City's Public Sphere

Introduction

Speaking about the 'city' is getting harder and harder. This is not just because contemporary cities have reached levels of extreme complexity, but because the notion of 'city' itself is facing a crisis never encountered before. Can a complex network of streets, buildings and green areas, of slums and glamorous commercial zones be considered a city just because it hosts a large number of people? This question has engaged an extremely large number of architects, planners and sociologists in trying to figure out how contemporary urban space is evolving, and how the instruments and the practice to deal with it should evolve as well. A common feeling of inadequacy has in fact become the base of a wide debate about the present and the future of cities, and many are questioning the efficiency of the existing approaches used to plan, design and manage a town:

> Having taken on board the theme of the crisis of planning, whether this is true or supposed, we have seen as useful, or even necessary, to verify whether the traditional planning instruments are still able to fulfil the needs of cities that are seeking their lost identity or looking for a brand new one (Lo Piccolo, 1995, p.19) [translation by author].

The next few pages review some of the crucial issues that relate to this crisis of identity of cities and uncertainty about the adequacy of planning. The main issues of fragmentation of space, society and political agency are examined together with their consequences on urban scenarios. Several tensions characterising contemporary urban life, and the way we analyse it, are highlighted. Among these, the tension existing between the 'global' aspects of contemporary society and economy, and the need for protecting local interests, cultures and communities, plays an important role, as do the gradual 'commodification' – the reduction to a commodity – of places often carried out by the 'global' agents and the consequent crisis of urban public space. The overall scenario in which these tensions have to be framed is a situation of conflict and transition between the centralised urban control, promoted by a declining modernist planning, and the fragmented post-modernist vision of the contemporary city.

For a few years, advocates of the social, public potential of computer networking, have argued that the application of new telecommunication technologies, and in particular the electronic 'space' better known as cyberspace, could play an important role in the restructuring and revitalisation of urban space, or at least of the public sphere. In the second part of this chapter, several different points of view about the relation between advanced telecommunications and the city are reviewed. This review suggests noticing how theoretical positions that can seem very different from each other, some seeing cyberspace as a solution to urban problems, others envisaging a major threat in the emergence of 'virtual' environments, are often flawed by a deterministic attitude. Nevertheless, many of the possible advantages and drawbacks of the application of cyberspace to cities would still have to be considered and taken on board, but as investigation themes addressing real innovation in contemporary cities, rather than as the products of forecasts of possible futures. In order to achieve a better understanding of this relationship between computer networking and the city, an important thing to do is to look not just at the 'product' as an IT system, but to look at the socio-technical processes that have been underpinning its development.

This whole book is based on the assumption that a more nuanced vision of the relationship between technology and society has to be adopted, coherently with the theoretical perspective known as 'Social Construction of Technology'. This type of approach does not necessarily contrast with the interesting, dystopian critique of the utopian and futurist positions made by many political economists, but it tends to integrate it with a more 'localised' vision, that acknowledges the influence of local actors as councils and firms, on the shaping of urban public cyberspace and its consequences on the city.

The crisis of contemporary cities

The major problem acknowledged by Lewis Mumford, in the 1960s, was 'the sprawling giantism of the twentieth-century city, that was leading inexorably to megalopolis and thence to necropolis, the death of the city' (Mumford, 1960; quoted in Wilson, 1995 p.147). In the contemporary scenario, though, we do not necessarily need to consider big metropolitan areas like Los Angeles or Mexico City to find signs of crises of identity and a sense of citizenship, as well as many other social problems linked with unemployment, multicultural and multiracial conditions, and a lack of communication among people. In other words, the critical state that Mumford was envisaging for the big American cities has extended its influence to many average size cities around the world. This is not necessarily a function of urban size. It is true that the growth of many metropolitan areas has radically changed the relationship between town and country, blurring or even dissolving the city limits. It is hard to perceive a city with no borders that moreover lacks a strong characterisation, because of the 'international style' of its suburbs, as a unique site, with its own culture and personality, capable of giving its inhabitants the sense of community and confidence they could need. But this is not the only cause for the apparent lack of definition of urban space. Many scholars find it hard to recognize a city not because of the absence of an edge, but because either the physical and the social space that were so far defined as 'town' have broken up into bits that seem independent and detached from each other:

It is often said that disappearance is the city's fate. But the 'city' has already disappeared, and for those like me who deal with urbanism, the problem is letting it come back. The city in which we live nowadays is not any more a city. It has many names: metropolitan conurbations, residential areas, industrial areas, commercial areas, holiday areas. Our life is organised by areas, but what we define a city has already been widely overtaken (Magnaghi, 1995, p.17) [translation by author].

Cities gone to pieces

So, the crisis is not just a matter of size. It is not only because the city, as a whole, has grown too much. Actually, it is also because the city is not a 'whole' any more or, perhaps, it is not like the majority of us expect a 'whole' to be. The borders between town and country, and even between different neighbouring towns are disappearing, yet new edges are developing inside the city itself.

'Fragmentation' is probably the keyword when it comes to speaking about contemporary urban space, and this phenomenon can be observed in social life as well as in the production of the built environment. A clear example of this is given by Christine Boyer speaking about recent developments in New York, and criticising an article written by the architectural critic Paul Goldberger, who

> ...hopes to convince us that architects and planners are scene-makers, not social reformers; they need no longer be concerned with the difference between public, semi-public or private space in the city; and they generally have lost the need or desire for social accountability. From this point of view, architecture in the city is an autonomous artistic expression that has nothing to do with the social and economic plight of the poor. I would argue that this view is part of the reason why our collective view of the city has become fragmented into individualized pieces incompatible with the creation of a physical plan, and why we have no map of the city linking together the poorer neighbourhoods with the enclaves of the well-to-do. A plan or a map might draw us closer together and underscore our collective plight (Boyer, 1993, pp.112-113).

These words are a strong critique of contemporary planning and urban design, and the acknowledgement that the city is actually becoming a place with no map or plan, being rather a series of places divided by social and cultural gaps between which any type of communication is getting more and more difficult to be held.

On one side there are the commodified and heavily controlled parts of the growing metropolis, such as shopping malls, office buildings, airports, entertainment sites. On the other side are those parts of the city that have problems in keeping their local and cultural characteristics, having to cope with deprivation at several levels, and find themselves isolated from the wealthy urban space.

Sorkin calls them the 'departicularized' space on the town, which is

> Obsessed with the point of production and the point of sale, the new city is little more than a swarm of urban bit jettisoning a physical view of the whole, sacrificing the idea of the city as the site of community and human connection (Sorkin, 1992, p.xiii).

This lack of connectivity is to be considered from multiple points of view. The crisis of social connections, and some major changes in the way urban space is designed and built, interact in a cause-effect circle, that transforms the city into a jigsaw of different pieces, sometimes indifferent to each other, yet contiguous.

The 'spatial' side of this process was acknowledged several years ago by a range of architectural theorists. It was noticed that connections and paths tended to become at most 'by-passes' from one zone to another, losing that characteristic of micro-universe where the features of the neighbourhood and the whole city were displayed to the visitor (Norberg-Schulz, 1971). It cannot be denied, then, that space and society are interconnected, and that the lack of social connectivity affects urban space, and vice versa.

As has been said before, the Los Angeles described by Mike Davis in his book City of Quartz (1990) offers a remarkable example of urban fragmentation of space and society:

> Welcome to post-liberal Los Angeles, where the defense of luxury lifestyles is translated into a proliferation of new repressions in space and movement, undergirded by the ubiquitous 'armed response'. This obsession with physical security systems, and, collaterally, with the architectural policing of social boundaries, has become a zeitgeist of urban restructuring, a master narrative in the emerging built environment of the 1990s (Davis, 1990, p.223).

But such trends can also be observed in relatively small European cities like Newcastle, where the differences among 'rich' and 'poor' areas have been increasing. The feeling of uneasiness for wealthy people and the lack of security that accompanies social polarisation have lead to the fact that parts of the city – and indeed many other cities in the UK – have to be controlled by public CCTV security systems (Graham and Marvin, 1996). The territory tends to be organised into zones more and more disconnected from each other. The new developments tend to be designed as 'closed systems'. The Newcastle Business Park, a big complex on the banks of the river Tyne, for example, is completely indifferent to the deprived area nearby. Rather, it protects itself from its surrounds with high fences and video surveillance. Another example is the central 'Eldon Square' shopping centre, a pseudo-public space that has replaced a remarkable portion of the Newcastle city centre. As most large malls, it has its private police, a strict surveillance and closes for the night. Some have observed that this process is generated by 'corporate and state planners, [who] have created environments that are based on desires for security rather than interaction, for entertainment rather than (perhaps divisive) politics' (D. Mitchell 1995, p.119). This need for security and the separation that derives from it, affect not only the characteristics of central business areas and shopping malls, but also increase hostility and suspicion among different residential areas, in which local communities tend to close themselves even more to the outcomers or to socially, culturally, economically different groups of people:

In areas of our cities where poverty and social diversity are concentrated, the signs of stress are pervasive, in the routinisation of violence, alienation and anger, of crime and stigma. This reinforces exclusionary tendencies, encouraging a defensive sense among the better-off, labelling these stressed neighbourhoods as 'outside', not part of the 'mainstream', and 'other' (Healey et al., 1995, p.7).

This change of scale, from the whole city to the particular community or even single individual, is both cause and effect of a major change in how urban space is managed. The fragmentation of space and society is strictly linked to a fragmentation of political agency, as described by Simin Davoudi:

City governments are no longer the key locus for integration of urban relationships, but merely one of many actors competing for access to resources and control of agendas (Davoudi, 1995, p.226).

The traditional forms of urban governance are weakened by this change of approach towards government in general. Neo-liberal tendencies, by promoting the concept of a 'lighter' public administration, have encouraged strategies of privatisation and deregulation. This has resulted in the 'offloading to companies, voluntary associations and households of tasks formerly undertaken by the state' (Healey et al., 1995, p.5) This privatisation process and its relationship with the changes in the social fabric of urban places is also described by Dear as follows:

At base, I believe that one of the most characteristic features of contemporary urbanism is fragmentation. This finds expression in (…) patronage driven development: opportunities opened up by the collapse of government oversight. Paradoxically, such fragmentation at the same time encourages the proliferation of intense local autonomies. Related to this is an emergent privatism. During the 1980s a reassertion of individual rights over community obligations has resulted in a loss of community (Dear, 1993, pp.7-8).

Such a claim underlines both the benefits and the drawbacks of this phenomenon. If, in fact, on one side, new opportunities for individuals and for certain minorities can be opened by an increased lack of central control, whilst on the other side it is evident that fragmentation can affect everything that was previously seen as a 'whole', at different scales. Local communities can, then, themselves live a growing condition of crisis, leaving space for the influence of well organised lobbies:

The instability and fragmentation of relations and networks represents threats to political community and allow powerful groups to impose their criteria and practices discretely and often unchallenged (Mabbot, 1993; quoted in Davoudi, 1995, p.27).

Towns as goods: globalisation and city marketing

Cities, then, have become a yet more complex arena, in which a decreased control over the individuals and marginalised groups has its counterpart in a remarkable influence on key decisions and town management by big corporate firms and private lobbies, that may have no real 'roots' in that specific reality. Cities are increasingly seen as places that can attract investment, rather than producing it endogenously. This, together with the lack of public funding and diminished public control, encourages the 'commodification' of places, the management of urban spaces as commodities that have to be 'chosen' and bought by consumers and investors:

> The perception of crisis is producing an explosion of interest in the city. Much of this is informed by the belief that the quality of a city can become an economic asset in the struggle to capture and retain mobile globalising company investment (Healey et al., 1995, p.7).

> As the economic fortunes of cities move to the top of the local political agenda, urban economic policy making, 'public private partnerships', and place marketing are emerging as driving forces in changing approaches to urban management. Central here is the desire to position cities favourably on the global economic networks dominated by the burgeoning multinational corporations (Graham, 1995, p.85).

This can, sometimes, bring no apparent benefits to the local people. Rather, the interests involved in treating the city as a commodity can be opposite to those of the local population, as stated by Enzo Mingione:

> The much expanded role of non-residents (e.g. businessmen, investors, tourists, shoppers, students, cultural visitors and commuters) complicates urban life and produces a new type of socio-economic environment, no longer dependent nearly exclusively on the number of permanent residents (Martinotti, 1993). It is on this ground that a fracture between tax-payers' interests (mostly residents and local businesses) and city marketing interests (mostly oriented towards capturing, at whatever the cost, increasing numbers of city users and regular business visitors) is widening fast and in some cases assuming greater importance than traditional class conflict (Mingione, 1995, p.197).

The crisis of public space

This conflict of interests between the city as a place and the city as a commodity, has some major effects on the parts of the town that used to belong to the community and that increasingly are used to produce money and to promote the image of the city to the outcomers. Elizabeth Wilson takes Paris as an example noting that:

> The Parc de la Villette is designed for tourists rather than for the hoarse-voiced, red-handed working men and women who in any case no longer work or live there. Thus it is in the great cities of the world at least, but also certainly in any smaller cities that can

capitalize on an historic past, or an industrial peculiarity – not only is the tourist becoming perhaps the most important kind of inhabitant, but we all become tourists in our own cities... (Wilson, 1995, p.157).

The public places of the contemporary town are increasingly being turned into lucrative areas, reflecting the interests of private capital. Because of this, they also have to be controllable. The comparison made by Wilson when she states that airports are 'Increasingly indistinguishable from shopping malls' (Wilson, 1995, p.157) can be reversed. If it is true that airports are being 'consumerised', being filled up with shops, downtown shopping malls resemble air terminals, as well, because of their strict surveillance that lacks only the metal detectors at the entrance.

Most of these 'post-urban' spaces, as Wilson defines them, are in fact rarely truly open zones. Access is carefully controlled and to be secure they actually need to be protected, closed areas with few connections and relationships with the city outside. They can be seen as a sort of modern medieval keeps, with surveillance cameras rather than walls, guarding the private properties of the new landlords. The difference with the middle ages is that outside public space is disappearing, the marketplace is becoming itself a fortress, and squares are what is among the buildings, rather than the buildings what surrounds squares. The 'death' – or the agony – of the city is then to be related to the crisis of its public, open bits. Therefore Christine Boyer defines the 'City of Illusion as calling something public space when indeed is not; focusing on the provision of luxury spaces within the centre of the city and ignoring most of the interstitial places' (Boyer, 1993, pp.113-114).

Out of sight

So, when we say that the city is 'disappearing', does it mean that it is dissolving and something else is replacing it, whether it is a big theme park or a landscape of fortresses? Are 'the connections that make sense of forms' (Sorkin, 1992, p.xiii) being destroyed? Or are they just hidden? Elizabeth Wilson seems to suggest this when she states that

we must live in this realm of smudged boundaries, of pollution and disorder, and the gentre of urban writing is beginning to recognize its romance. It was always the interstices of the city, the forgotten bits between, the corners of the city that somehow escaped, that constituted its charm, forgotten squares, canals, deserted houses – private, secret angles of the vast public space (Wilson, 1995, p.160).

Words like 'ignored' or 'forgotten' actually suggest that, although the commodification of public space is a real fact, a very important cause of the crisis of contemporary cities is invisibility, and the related lack of communication between people. This phenomenon is itself deeply linked with the privatisation processes that are going on. As Sorkin notes, when he compares post-modern, commercialised spaces to theme parks:

The theme park presents its happy regulated vision of pleasure – all those artfully hoodwinking forms – as a substitute for the democratic public realm, and it does so appealingly by stripping troubled urbanity of its sting, of the presence of the poor, of crime, of dirt, of work. In the 'public' spaces of the theme park or the shopping mall, speech itself is restricted: there are no demonstrations in Disneyland (Sorkin, 1992, p.5).

This crisis of visibility has two different aspects that can be grasped from the previous quotations. The real city, and society, tends to be invisible in the highly controlled and commodified places. At the same time, these wealthy bits of urban space tend to marginalise the other areas, both economically – they absorb most of the investment – and socially. Public spaces and architectures, built with the purpose of gathering people, lose their attractiveness because of lack of safety, because they have no appealing features left, or simply because they have become invisible, forgotten, and nobody goes there and uses them. The times in which urban design and architecture were seen as capable of benefiting society as a whole, seem definitely gone.

Beyond modernity

Curing the illnesses of the fragmented urban space by building something new seems too difficult for architects and urban designers, and it can be even cause for further fragmentation. We are in a very particular phase of history, in which architects feel like losing that creative power that was believed to make them capable to modify the environment positively, making people's existence better through the use of materials, textures, shapes.

A sort of uneasiness is felt for all the disciplines that imply actual intervention on urban space, not excluding planning. It is, in a way, the fall of the modernist ideals of rational management of space. In 1929, Le Corbusier, one of the masters of modernist architecture and planning wrote that

A town is pure geometry. When man is free, his tendency is towards pure geometry. It is then that he achieves what we call order. Order is indispensable to him, otherwise his actions would be without coherence and could lead nowhere. And to it he brings that aid of his idea of perfection. The more this order is an exact one, the more happy he is, the more secure he feels (Le Corbusier, 1929, pp.28-29).

It is this dream of perfection, of harmony, transparency, and visibility achieved through the use of pure geometrical entities such as the right angle, together with the rational use of light and materials. It is the modernist determinism based on a 'overflowing confidence in science and technology' (Madani-Pour, 1995, p.22) that is supposed to overcome chaos, producing a perfect, ideal, untouched city-machine that affects people's lives:

Then suddenly we find ourselves at the feet of the first sky-scrapers. But here we have, not the meagre shaft of sunlight which so faintly illumines the dismal streets of New York, but an immensity of space. The whole city is a Park. The terraces stretch out over lawns and into groves. Low buildings of a horizontal kind lead the eye on to the foliage of the trees. Where are now the trivial Procuracies? Here is the CITY with its crowds living in peace and pure air, where noise is smothered under the foliage of green trees. The chaos of New York is overcome. Here, bathed in light, stands the modern city (Le Corbusier, 1929, p.177).

But dividing and zoning cities, as well as designing white, hyper-hygienic, well-lit habitation, based on 'pure' anthropomorphic dimensions, was seldom, or never, possible in the ideal way imagined by Le Corbusier. Its partial application anyway did not improve society and quality of life as it was supposed to do. Rather, it is now believed that urban space developed its fragmentation possibly also because of this too idealistic approach. Barry Goodchild seems to suggest it when he says:

> The big city continues to be seen as a place of excitement and innovation and also a place of danger, pollution and social segregation. Yet planning is now so deeply implicated in the creation of the city in the twentieth century, that the earlier hopes of town planning as an exercise in social reform seem deeply flawed (Goodchild, 1990, p.132).

Are, then, town planning and architecture guilty of having produced the crisis of contemporary urban space? What we can say is that the way society, culture and economy have changed – and are changing – has ruled out any type of idealistic solution for town development and management. As Dear argues:

> For me, postmodernity encompasses a set of perspectives on a most impressive intellectual dilemma: that Rationalism has failed both as an ideal and as a practical guide for thought and social action; and that henceforth, we have to manage without such Enlightenment desiderata as decisive theoretical argument or self-evident truth. In Toulmin's (1990) felicitous phrases, we have moved from a mindset which emphasises the written, the general, the universal and the timeless to one which privileges the oral, the local, particular and timely (Dear, 1995, p.28).

The postmodern city has shown clearly that real life is somewhat different and not very compatible with the tidy, mechanical world sought by the modernist architects and planners. Koolhaas and Mau also directly criticise current planning practices as inadequate to pro-actively and strategically address the problems of contemporary cities:

> Dissatisfaction with the contemporary city has not led to the development of a credible alternative; it has, on the contrary, inspired only more refined ways of articulating dissatisfaction. A profession persists in its fantasies, its ideology, its pretension, its illusions of involvement and control, and is therefore incapable of conceiving new modernities, partial interventions, strategic realignments (Koolhaas and Mau, 1995, p.965).

Parallel to these calls for a redefinition of place-making and managing practice, technology has been the object of enthusiastic approaches that tend to see its new developments as the long-awaited solutions for the problems of the fragmented urban space. The advent of information technology and advanced telecommunications systems, has seemed to be opening new perspectives for the management of cities and public space, encouraging a renewed deterministic and rationalistic approach by some scholars. New threats and old drawbacks, according to some critics, seem to emerge, as well. The next sections will deal with these different positions.

Cyberspace as a new world

One of the possible choices, when faced with any of the dimensions of the 'urban crisis', can be to leave physicality behind and to look for perfect or quasi-perfect 'environments'. It is not the first time in which city-makers have to develop their ideas in unreal territories, trying to avoid the limits imposed by the physical world and the complexity of society. The history of architecture and urbanism is full of examples of virtual cities and virtual buildings represented on paintings and sculptures, or simply on drafts and designs that never become real. Marcos Novak has given just some of the possible examples saying that

> Piranesi's series of etchings entitled Carceri, or Prisons, marks the beginning of an architectural discourse of the purposefully unbuildable. Against the increasing constriction of architectural practice, Piranesi drew an imagined world of complex, evocative architecture. (...) Ledoux, Lequeu, Boullèe, each contributed to this struggle: as architectural practice was made more and more prosaic by the encroachment of utility, they responded by inventing a bolder and bolder imaginary counterpart (Novak, 1991, p.246).

Visionary architecture is indeed a relevant and interesting bit of history, and it has definitely influenced, as an avant-garde and a cultural sign of the times, the actual production and conception of the built environment and urban culture.

But there are big differences between the virtual worlds known and used by our ancestors and some of those available now. Information technology has in fact made possible the creation of 'virtual worlds' that are not just ways to represent an alternative, ideal environment, and communicate its characteristics to an audience. Through computer-based virtual reality, and the application of telematics, it is actually possible to do more than just display something: it is possible to interact and establish two-way communication with both the objects and the people who share a certain electronic environment. In other words, these virtual worlds work and have functions that go beyond mere representation, whether we are speaking about world-wide data networks or local and more limited experiments.

Telematics is not actually something new. The 'art' of networking computers using modems and telephone lines, or better hardware, has been practised for several decades by professionals and amateurs, establishing services and communities very similar to those existing among radio-amateurs. But the current success of telematics and the reason why everyone has been talking about it is that it has a name: Internet. Born as a military network, developed for academic purposes, the Internet is now a 'network of networks' embracing academic, business and private users all over the world. Statistics show that in October 1994 the number of users capable to use all the facilities provided by the Internet was 13.5 million, while 27.5 million people were able to use the basic e-mail service to communicate. In just three years (1997) these figures had increased to 57 million Internet users and 71 million people with email access (statistics by the Matrix Information and Directory Service – MIDS). In January 2003 an 'Internet Domain Survey' from the Internet Software Consortium (www.isc.org) could count over 171 million active 'hosts', that is computers actively connected to the global network. This figure probably does not take into account most of those machines, and households/people behind them, still connecting on a very part-time basis by dialing up their service provider. Each machine is very likely to serve more than one user, so the figures add up to an impressive growth rate.

The Internet, and more generally speaking the so-called 'cyberspace', has then been increasingly seen by many as an alternative environment in which to experiment with virtual spaces that enable people to communicate among them and perform several functions previously bound exclusively to the physical world.

The promised land of cyberspace: utopia and futurism

So, cyberspace seems to embody both the characteristics of real and virtual space. It 'works', enabling people to do real things and it has not, at the same time, all the disadvantages of the material world. As Katherine Hayles states: 'Being able to occupy a virtual space implies that one can have the benefits of physicality without being bound by its limitations' (Hayles, 1993, p.179). Cyberspace looks like the place where it can be possible to 'build' the ideal, rationalised city which modernism failed to produce for real, and this can sound really appealing for those architects, urban designers and planners who have been looking at high technology as the instrument to overcome all the problems of communication and fragmentation – both social and physical – that affect contemporary cities. Cyberspace is here, and from an utopianist perspective it has been claimed to be

> The realm of pure information, filling like a lake, siphoning the jangle of messages transfiguring the physical world, decontaminating the natural and urban landscapes, redeeming them, saving them from the chain-dragging bulldozers of the paper industry, from the diesel smoke of courier and post-office trucks, from jet fuel fumes and clogged airports, from billboards, trashy and pretentious architecture, hour-long freeway commutes, ticket lines, and choked subways... from all the inefficiencies, pollution (chemical and informational), and corruptions attendant to the process of moving information attached to things – from paper to brains – across, over and under the vast and bumpy surface of the earth rather than letting it fly free in the soft hail of electrons that is cyberspace (Benedikt, 1991, p.3).

This is not just the definition of a technique, a tool that should be used because it provides alternative ways to manage cities and their functions. This sounds almost like a new chance for salvation, something that will change the way things are so dramatically and so positively that it deserves to be described with a verb like 'redeeming'. It is actually a triumph of idealism, fed by the positivist, techno-enthusiastic attitude that telematics could be the ultimate solution for urban problems. Cyberspace has been seen by architects like Michael Benedikt as a chance to revitalise architecture and urbanism, that type of creative process that has the power to generate spaces where it is good to live for everyone, the ability to turn cities into clean, unpolluted, citizen-friendly environments, both in electronic and physical dimensions. Benedikt stressed a concept of cyberspace as a dimension which transfigures the physical world, because it facilitates a process in which 'architecture shifts from the traditional continuous city to a structure of relationships, connections and associations' (Benedikt, 1991, p.249). Marcos Novak went even further, drawing a parallel between virtual space and his discipline saying that: 'cyberspace is architecture; cyberspace has an architecture; and cyberspace contains architecture' (Novak, 1991, p.226). The electronic environments can be, in a way, the actualisation of what before was considered just visionarism:

> It is instructive to scan the manifestos (of visionary architecture) for premonitions of an architecture of cyberspace. Many have contributed to this effort, becoming the world's front line of imagination, building in words and images what we can not yet convince the physical world to bear (Novak, 1991, p.246).

And such commentators seem to go well beyond visionarism, because these new worlds are not mere representations: they have functions, people, interaction, and therefore they can host a new type of architecture that deserves its name:

> A liquid architecture in cyberspace is clearly a dematerialized architecture. It is an architecture that is no longer satisfied with only space and form and light and all the aspects of the real world. It is an architecture of fluctuating relations between abstract elements. It is an architecture that tends to music (Novak, 1991, p.251).

But the cyber-enthusiasts cannot be found only among architects. The hype for new technologies is extremely strong in nearly all the fields of human activity. Michael Ogden deals with politics, but his hopes for the future are not less optimistic than those of Benedikt and Novak:

> Cyberspace is a conceptual 'spaceless place' where words, human relationships, data, wealth, status and power are made manifest by people using computer-mediated communications technology. It has been variously described as a new universe, a parallel universe created and sustained by the world's present and future computers and communication networks. A world in which the global traffic of data and knowledge, facts and lies, entertainment and alter-human agency take on form with tens of millions of voices and twice as many eyes in a silent, invisible concert of inquiry, deal making, dream sharing, and simple beholding. It is accessed through any computer capable of linking into the network; a place, one place, multidimensional, limitless, everywhere and nowhere, a place where nothing is forgotten and yet everything changes (Ogden, 1994, p.715).

This new society, where the opportunities are endless and for everyone, as described by Ogden, belongs to what is defined as the 'anything, anytime, anywhere' futurist myth on the positive impacts of cyberspace (Graham and Marvin, 1996, p.88). It is the same idea of overcoming the problems of physicality that architects have projected onto politics, social affairs and culture: 'We're creating a womb for culture at large, for civilisation, and I think that's a potentially beautiful thing' (Rushkoff, from *Once upon a Time in Cyberville*, Channel 4 Television, 1994). Being able to get rid of distances and of the oppressive habits inherited by the industrial society, seems to some writers the key for a new, inevitable era, in which displacement wins over agglomeration. Cities are then seen as an inconvenient heritage that has to change radically, eventually to dematerialise into virtual space. That is the era of the 'electronic cottage', the 'wired' home that becomes workplace strengthening family and community bonds:

> If the electronic cottage were to spread, a chain of consequences of great importance would flow through society (...) Community impact: (...) The electronic cottage could help restore a sense of community belonging, and touch off a renaissance among voluntary organizations like churches, women's groups, lodges, athletic and youth organizations. (...) Environmental impact: (...) Instead of requiring highly concentrated amounts of energy in a few high-rise offices or sprawling factory complexes, and therefore requiring highly centralized energy generation, the electronic cottage system would spread out energy demand and thus make it easier to use solar, wind and other alternative energy technologies (...) Economic impact: (...) if individuals came to own their own electronic terminals and equipment, purchased perhaps on credit, they would become, in effect, independent enterpeneurs rather than classical employees – meaning, as it were, increased ownership of the 'means of production' by the worker. (...) Psychological impact: (...) work at home suggests a deepening of face-to-face and emotional relationships in both the home and the neighbourhood. Rather than a world of purely vicarious human relationships, with an electric screen interposed between the individual and the rest of humanity, as imagined in many science fiction stories, one can postulate a world divided into two sets of human relationships – one real, the other vicarious – with different rules and roles in each (Toffler, 1981, pp.214-216).

People were waiting for the end of the world in the year 1000, and towards the end of the second millennium many were claiming that another brand new world was going to replace or at least improve our reality. This optimistic and technocratic vision has not just been promoted by utopian scientists and artists. Industry has had a great deal of self-interest, for obvious reasons, in encouraging such an enthusiastic attitude. It is not by chance that Bill Gates, chairman of Microsoft Corporation, has been speaking just the same language as those intellectuals who tend to blindly support the idea of a perfect relationship between the physical world and cyberspace. Asked in *The Guardian* about possible losses of jobs because of the advent of information technology, he stated:

> Until we have infinitely good education, infinitely good service everywhere and beautiful inner cities, there are jobs left to be done. The day that most important needs

are met, society will shorten the work week and increase vacation time. As the world gets richer, some of its wealth will go to increasing relaxation time (Gates, 1995).

Thus, dominant discourses about cyberspace tend to be a mixture of cyber-philosophy and information technology commercials. These two points of view end up being the two sides of the same coin. Techno-intellectuals encourage people to buy technology and services, to think about them as a way to get empowered. So do Microsoft, IBM, AOL etc.

No wonder that the cover of the first British issue of Wired magazine solemnly stated: 'We have it in our power to begin the world over again', and this creative ability was claimed to be equally shared among virtually every person who would not retreat from the flow of progress. *Being Digital* (1995) is the title of a book written by Nicholas Negroponte, guru of technology and chief of M.I.T. Media Lab., but it has also been an imperative given by all the media to those people who would not want their way of life to become obsolete. The status-symbol gadgets of nowadays are objects allowing treatment of information itself. Having access to broadband, databases and 'portals' with all the latest information, being able to tap into these archives even while you are one the move, through several types of wireless mobile devices, whether you really need this or not, and even if you cannot manage the sheer amount of information, is portrayed and perceived as what makes the difference. And, of course, industry has been offering all of this, information and the objects you need to deal with it, like state-of-the-art 'personal digital assistants' (PDAs) and second and third generation mobile phones that represent a convergence between mobile computing and mobile telephony, wireless yet in touch with the whole world. Being able to displace ourselves working from a park bench as in a Microsoft's commercial from the Nineties, was presented as just the beginning of a new era of freedom from the constrictions of physical world.

Techno-determinist dystopia

In such a discourse, the parallel reality of cyberspace has been seen as somehow inevitable, and it is argued that all of us spend a remarkable part of our time dealing with virtual environments of some kind, both for business and entertainment: 'We are approaching an epoch within which a self-identity derived from "real" "embodied" experiences is unable to compete with ones derived from the "erotic ontology" of hyperreal simulated disembodied cyberspace' (Burrows, 1995, p.8). Groups and schools of thought were born and grown up to cope with the new forthcoming robotic era. Timothy Leary in 1990 wrote on Pataphysics Magazine:

> We are mutating into another species, from Aquaria to the Terrarium, and now we are moving into Cyberia. We are creatures crawling to the center of the cybernetic world. But cybernetics are the stuff of which the world is made. Matter is simply frozen information... (Leary, 1990; quoted in Leary, 1994, p.vii).

A very common attitude in dealing with technological innovation, that is highlighted in Leary's statement as well as in the futuristic claims that were examined before, is determinism. Technology, and in particular cyberspace, is seen as the primary cause for an evolutionary process that brings a whole series of dramatic consequences to physical space and society:

> To technological determinists, the urban impacts of telematics and telecommunications are relatively straightforward, linear, direct and easy to pinpoint. Most are virtually inevitable. The decentralisation, or even dissolution, of cities; the free availability of highly capable communications in all locations; the shift towards city economies based on information; the growth of a culture based on tele-interactions; the shift to an 'immaterial' urban life; the growth of telecommuting – all will be shaped by new innovations in telematics in a deterministic and inevitable fashion. The analytical and policy issues suggested by technological determinist therefore centre around how society can adapt to and learn to live with the effects of telecommunications-based change, rather than focusing on the ways in which these effects may be altered of reshaped through policy initiatives (Graham and Marvin, 1996, p.83).

Techno-determinists are, then, concerned with the impacts of telematics and cyberspace, not with their shaping. Their efforts are focused on how society, and even single individuals, should learn to adapt to the new forthcoming condition, to mutate, to use Leary's evolutionary expression.

And technological impacts on society are not necessarily to be seen as positive as Gates, Ogden, Benedikt and many others see them. Dystopian views are as strong and categorical as utopian ones. They have been depicting a future in which people's freedom is diminished or put in jeopardy by advanced telecommunications. Cyberpunk, for instance, is a literary and cultural movement characterised by a not very optimistic vision of our technological – and social – futures. Cyberpunk science fiction books like the ones written by William Gibson or Bruce Sterling in the 1990s, depict a world in which people live in a completely commodified hi-tech world, heavily ruled and controlled, in which inequality is sovereign. This literature provides: 'stunning examples of how realist "extrapolative" science fiction can operate as prefigurative social theory, as well as an anticipatory opposition politics to the cyber-fascism lurking over the next horizon' (Davis, 1992 – quoted in Burrows, 1995, p.2). In Gibson's novel *Virtual Light* (1993), the only alternative available against a society based on lies and social control through video surveillance and telecommunication networks, seems living in the interstices, in the marginalised slums of the city that scare the well-to-do people so much that some of them get to believe that those places actually host ghosts and cannibals:

> ...And that bridge, man, that's one evil motherfucking place. Those people anarchists, antichrists, cannibal motherfuckers out there, man... 'I heard it was just a bunch of homeless people' Rydel said, vaguely recollecting some documentary he'd seen in Knoxville, 'just sort of making do'. 'No, man' Freddie said, 'homeless fuckers, they're on the street. Those bridge motherfuckers, they're like king-hell satanists and shit. You think that you can just move on out there yourself? No fucking way. They'll just let their own kind, see? Like a cult...' (Gibson, 1993, p.139).

According to the cyberpunk movement, future society is going to be even more fragmented, continuing the processes that are in act at the moment, and information technology is not inverting the trend, rather it is boosting it. In this hopeless scenario, the problem is not any more 'putting the pieces of society and cities together', as that is impossible to do. Rather, technology is useful as a way to exploit fragmentation, being able to go – most of the times virtually – where others cannot go, and being able to see what others cannot see. Cyberspace is therefore a parallel but rather separate domain from the 'real' world, a new frontier that still provides degrees of freedom and chances to change the course of events, for those capable of interacting with the electronic dimension. Power in cyberspace is in the hands of big corporate companies that rule the world as well as of a group of brave, outlaw hackers who have learned to get advantages from such a society, and resist against the dominating institutions. It is a dystopian point of view, and it tends to be deterministic as well, as it envisages the unstoppable advent of a whole social system. The process of shaping technologies is one-way, and leaves no space for local political and social agency. The impacts cannot be changed, at most people can learn how to survive, adapt and eventually exploit something that was imposed to them.

Beyond the cyberpunks, a world of annihilation, rather than of empowerment is envisaged by the philosopher and planner Paul Virilio, who sees in the abolition of space and time, allowed by telematics, the disappearance of individual agency:

> What remains to be abolished – and urgently – can only be space and time. (…) At the end of the century not much will remain of this planet that is not only polluted and impoverished, but also shrunken and reduced to nothing by the technologies of generalised interactivity (Virilio, 1993, p.12).

Virilio has also denounced the political threats that the development of cyberspace brings with it. To Virilio, virtual worlds can be the ultimate tool for globalisation. This, together with the fact that they 'possess an incomparable power for suggestion', makes the advent of new tyrannies possible:

> On leaving the White House in 1961, General Eisenhower declared that the military-industrial complex was a 'threat to democracy'. He knew what he was talking about, having set it up in the first place. In 1995, with the establishment of a real informational-industrial complex, and as various US politicians, notably Ross Perot and Newt Gingrich, talk of 'virtual democracy' in a tone which echoes fundamentalist mysticism, how can we miss the warnings? How can we fail to see the danger of a real cybernetics of the socio-political sphere? (Virilio, 1995, p.4).

Cyber-capitalism and the political economists

Virilio's quotations have introduced us to a dystopian vision of a society in which telecommunication and information technology are the means to keeping or getting 'global' power. Several critics, most of them dealing with the historical-materialist tradition, focus their attention on the ways in which capitalism is changing, upgrading itself to new conditions, and basically keeping on boosting and exploiting inequalities. Far from willing to celebrate any type of saga of future cyber-heroes, whether they might live in a free society or in a noir, fragmented and dangerous city, Marxist researchers and political economists see new technologies primarily as the upgrade of instruments to control society:

> Changes in mobility and communication infrastructure and patterns, therefore, are not neutral processes in the light of given or changing technological-logistical conditions and capabilities, but necessary elements in the struggle for maintaining, changing or consolidating social power (Swyngedouw, 1993, p.305).

The political economic thinking of the Marxist researchers cannot then be ignored or underestimated. It constitutes an alternative view that, although it leaves few hopes for the future, focuses on the actual political-economic processes that are going on underneath the glossy cover of the technological hype, wherever it comes from. Their critique of the techno-enthusiastic positions of utopians and futurists has to be taken on board to avoid forgetting the influence that the macro-structures of society have on the shaping of technology and, more specifically, cyberspace:

> In the zealous imagination of 'information age' ideologues, once political impediments are removed from communication technology's revolutionary thrust, its ability to bypass traditional time and space boundaries promises a cornucopia of comforts and conveniences. This would-be 'depolitization' of technology and economics is, rather transparently, itself a politically charged construction of reality that attempts to disassociate communication/information institutions and industries from the human actors and beneficiaries involved in their design and development. In the technocratic model, it is not the who that matters, it is the how to, the technical processes downplaying the explicit social applications (Sussman and Lent, 1991; Quoted in Graham and Marvin, 1996, pp.101-102).

So, the 'global village' in which everybody is able to enjoy the empowerment deriving from virtual space can be seen also as:

> ...a rapidly changing economic space determined by economic units whose size and transnationality places them above social pressures and political controls... [this] ...is a tendency that, favoured by the internationalization process and by high tech, attempts to impose the abstraction of a space of strategic decisions over the experience of place-based activities, cultures and politics (Castells 1985, p.31).

Exploring the relationship of cities and technologies

So, together with the hopes for a society made better, fairer and more efficient by cyberspace, the literature about new technologies presents fears of inverse effects. Is cyberspace going to solve problems or is it going to increase them? Is fragmentation going to turn into integration thanks to the opportunities and visibility that cyberspace provides? Or is a cyberspace-mediated society tending to be even more polarised and unequal through a 'new model of economic accumulation, social organization, and political legitimation' (Castells, 1985, p.24)? Can local people find an environment that supports individualities and gives importance to peculiarities or will they dissolve in a global circus, where only powerful companies have voice? And what, consequently, will happen to public space and to 'public cyberspace'? Is democracy and communication to be boosted by virtual environments, or is information just condemned to commodification? Is cyberspace going to benefit real urban places, by giving them an extra dimension and making them work better, or will it be the negation of physical space, the detached world in which to escape from reality?

The answers to these questions, whatever they could be, tend at present to come just from speculation, rather than from proper research. It is a fact that this embryonic phase of the development of the 'fifth dimension' of cyberspace is characterised by futurologists' claims of a generalised and all-encompassing kind, that unfortunately tend to underestimate or even totally ignore the need for research on the phenomenon as it is now, and as it is being shaped and developed.

There is need for social research in this field, also because of a widespread tendency towards determinist positions, coming from both the techno-enthusiast and the techno-scared. Determinism is probably the main weakness of both of these points of view. Whether they are optimistic or pessimistic about new technologies and society, they still share an attitude that makes them scrutinise the future, yet quite indifferent about present times. What is going on now is seen as a mere series of facts, something that will affect future but that cannot be affected, modified, or carried out in a different manner. Consequently, the message lying within this type of analysis is a sort of invitation not to be unprepared for the new oncoming electronic era. We must 'be digital' as soon as possible, eventually buying the necessary equipment, because the new arena for competition, whatever type of competition it is, will be cyberspace. Even if this can be a good piece of advice at an individual level, and getting to grips personally with information technology is possibly a good thing to do nowadays, the problem is that this attitude shows all its limits when it comes to dealing with society and cities.

Political economists go beyond futurology of any kind, concentrating on the present processes that are affecting and restructuring society. But their vision does not leave many chances for local actors to affect these developments, as Graham and Marvin note:

> Political economy often overplays the conservative effects of the structures of capitalism in shaping technology and neglects the degree to which social processes can change telecommunications development. It often ascribes simple and all-encompassing powers to abstract and macro-level capitalist structures, while neglecting the ways in which structures are themselves created by innumerable individual and institutional actions over time (Graham and Marvin, 1996, pp.112-113).

The questions to be asked, then, deal with the shaping of urban cyberspace, primarily at the urban level. How are the agencies in cities developing their own virtual places, how are they being shaped and what are the consequences of this process, in terms of provision of information, social access, and relationship between physical and virtual city? These are not matters of fact that will lead to a future to talk about. These are processes going on or not yet started, that can be significantly affected by policies and choices, both made by public administrators as well as by citizens themselves and by industry's influence, in ways the political economists recognize. These processes, as we will see later on in this chapter, are also deeply affected by the 'visions' and the adoption of specific paradigms that underpin their existence.

It is not, for instance, a matter of fact that cities will heal themselves by 'jumping' into the perfect world of virtuality. Katherine Hayles warns about this kind of escape when she claims that '...leaving the body behind equates to the belief that if the problems won't go away from us, perhaps we can go away from the problems. Is it necessary to insist that nothing could be further from the truth?' (Hayles, 1993, p.183).

But we cannot as well take on board the catastrophic vision of an isolated '...individual who has lost, along with his or her natural mobility, any immediate means of intervening in the environment' (Virilio, 1993, p.11) without checking what is actually going on in terms of urban cyberspace, and considering any sensible alternative application of new technologies that could consider carefully the relationship with the physical world and a whole range of opportunities.

It is not, furthermore, a matter of fact that cyberspace is necessarily going to be a 'realm of pure information' that overcomes the chaos of daily physical life. As Julian Stallabrass notes: 'when and if this space really does become as public and immediate as television, what is to say that it will not become as cluttered and full of commercial garbage as anywhere else?' (Stallabrass, 1995, p.18)

Indeed, the more Internet usage grows, the harder it becomes to conceive cyberspace as an entity to be studied as something alternative to society – and detached from it. Bolter and Grusin for instance note that

> In matters pertaining to the theology of cyberspace, we must declare ourselves agnostics. We do not believe that cyberspace is an immaterial world, but that it is very much a part of our contemporary world and that it is constituted through a series of remediations. As a digital network, cyberspace remediates the electric communications networks of the past 150 years, the telegraph and the telephone; as virtual reality, it remediates the visual spaces of painting, film, and television; and as a social space, it remediates such historical places as cities and parks, and such nonplaces as theme parks and shopping malls. Like other contemporary mediated spaces, cyberspace refashions and extends earlier media, which are themselves embedded in material and social environments (Bolter and Grusin, 1999, pp.182, 183).

These observations call for research that is focused on the present and on recent history, rather than on speculation on possible urban futures. Understanding urban technology is possible only through the lens of contemporary times and the social, political, economic, artistic and cultural processes that are embedded in the development and usage of cyberspace itself.

Technology is not just a fact

In sum, all the theories and predictions about our digital futures are affected basically by the same problem: on the one hand they try to describe as inevitable certain aspects of future technologies that are still to be shaped. On the other hand what has been shaped, and the Internet for instance is more and more becoming part of the fabric of everyday life, is getting so ubiquitous that it ends up being taken for granted and not investigated and analysed. Research needs to be constantly carried out, to know how urban technology is evolving, what type of choices are driving its development, many of them related more to actual physical space and society than to the self-contained cyberworlds. Enthusiastic or not, the idea that underpins the determinist positions is about technology developing itself, with its own momentum, its own character, its own logic. Technology is seen as an independent factor that has its own shape, and the only thing that makes a difference is whether we decide to use it and how. But Guthrie and Dutton note that

> Like policy, technology is a social construction. (...) However, in the case of technology, these policy choices too often are obscured or overlooked because people focus only on decisions about the adoption or non adoption of a technology rather than also attending to decisions about design and implementation of the technology that influence its use and impact (Guthrie and Dutton, 1992, p.575).

The massive amount of hype about the information era makes people question how technology will affect society, politics, urban life. But this way to see things is flawed by the fact that technology itself is deeply affected by social, political, economic factors. 'The idea of a 'pure' technology is nonsense. Technologies always embody compromise' (Bijker and Law, 1992, p.3). This means that we must be aware of a two-way relationship between society and technology, and stop considering the latter as an independent variable. To borrow some concepts from Bijker and Law, it could be said that cyberspace

> ...might have been otherwise: this is the key to our interest and concern with technologies. Technologies do not, we suggest, evolve under the impetus of some necessary inner technological or scientific logic. They are not possessed of an inherent momentum. If they evolve or change, it is because they have been pressed into that shape. But the question then becomes: why did they actually take the form that they did? This is a question that can be broken down into a range of further questions. Why did the designers think in this way rather than that? What assumptions did the engineers, or the business people, or the politicians, make about the kinds of roles that people – or indeed machines – might play in the brave new worlds they sought to design and assemble? (Bijker and Law, 1992, p.3).

The importance of the early stages: change vs stabilisation

These remarks suggest that an investigation of technologies, and in particular of innovative technologies at an early stage of development, should consider very seriously what is behind the technology itself, what choices are being made, who is making these and of course why certain choices are prevailing. This consideration is crucial especially for the early phases of design, development and deployment of technological innovation, as brilliantly explained by Langdon Winner:

> Consciously or not, deliberately or inadvertently, societies choose structures for technologies that influence how people are going to work, communicate, travel, consume, and so forth over a very long time. In the processes by which structuring decisions are made, different people are differently situated and possess unequal degrees of power, as well as unequal levels of awareness. By far the greatest latitude of choice exists the very first time a particular instrument, system, or technique is introduced. Because choices tend to become strongly fixed in material equipment, economic investment, and social habit, the original flexibility vanishes for all practical purposes once the initial commitments are made. In that sense technological innovations are similar to legislative acts or political foundings that establish a framework for public order that will endure over many generations. For that reason, the same careful attention one would give to the rules, roles, and relationships of politics must also be given to such things as the building of highways, the creation of television networks, and the tailoring of seemingly insignificant features on new machines. The issues that divide or unite people in society are settled not only in the institutions and practices of politics proper, but also, and less obviously, in tangible arrangements of steel and concrete, wires and transistors, nuts and bolts (Winner, 1985, pp.30-31).

So, the best time to study innovation in urban cyberspace technologies, as indeed in any other technology, seems to be the early stages, as this is the time when characteristics and trajectories are particularly flexible, and tend later on to get 'stabilised' and fixed in an adopted, and much less changeable, form. It is then and there that meanings, specific visions and interpretations of society and reality, that come of course from specific social actors, are embedded into artefacts or initiatives, and influence their future configuration and use. The phenomenon of digital cities seems to be therefore ideally positioned, in chronological terms, to be studied under this perspective.

Technology as a process of social construction

But how could urban technology be looked at, if social factors and visions have to be taken into account? The sociological approach known as Social Construction of Technology (SCOT) provides a very effective framework that can be used to investigate processes of design, creation, deployment and stabilisation of technologies, and seems very suitable for multi-disciplinary studies. Wiebe Bijker explains that

> In the SCOT descriptive model, relevant social groups are the key starting point. Technological artefacts do not exist without the social interactions within and among social groups. The design details of artefacts are described by focusing on the problems and solutions that those relevant social groups have with respect to the artefact. Thus, increasing and decreasing degrees of stabilisation of the artefact can be traced. A crucial concept in SCOT (as well as in the Empirical Program of Relativism, EPOR, in the

sociology of scientific knowledge, to which SCOT is closely related) is interpretative flexibility. The interpretative flexibility of an artefact can be demonstrated by showing how, for different social groups, the artefact presents itself as essentially different artefacts (Bijker, 1992, pp.75-76).

Social constructivism has been often used within historical analysis of the development of technologies, for which it is a very powerful approach. Narratives on the social construction of bicycles (Pinch and Bijker, 1987), refrigerators (Schwartz Cowan, 1985), or weapons (Fallows, 1985), just to bring a few examples, have been brilliantly presented, and serve as an inspiration for knowing more about the technological developments of other artefacts and systems.

It can be argued, however, that a social constructivist perspective, or at least using elements from this approach, can be extremely effective and enlightening – though possibly harder to carry out – in real-time scenarios, when a new development is just going to happen. The insight that these kind of studies can provide could indeed help in explaining technological development as it happens, and above all draw suggestions on ways to improve the processes that are at the base of the establishment of a determinate technological system.

Interpreting and 'problematising' technology

So, in approaching the investigation of new technological implementations in cities, it is beneficial to look not just at the system itself, but at what generates the need(s) for it, the problems it is supposed to solve, the solutions it is supposed to provide, and whom and what visions are promoting a certain setup, or a set of alternative ones: 'Having identified the relevant social groups for a certain artefact, we are especially interested in the problems each group has with respect to that artefact. Around each problem, several variants of solution can be identified' (Pinch and Bijker, 1987, p.35)

As Pinch and Bijker argue, different ways of seeing – or interpreting – reality, create different perceived problems, as well as seeked solutions, and eventually different 'versions' and competing interpretations, and indeed different actual configurations, of the same technological object:

> We think that our account – in which the different interpretations by social groups of the content of artefacts lead by means of different chains of problems and solutions to different further developments – involves the content of the artefact itself (Pinch and Bijker, 1987, p.42).

Obviously these problems are – or can be – social and relational, not merely technological or functional. So, the social network or actors involved in the establishment of a technology, with the 'technological entrepreneurs' who promote certain 'solutions' in the first place, have to come to terms with their own roles and social, economic, political benefits they could derive from an initiative. The social dynamics behind the technological deployment are most important:

The contributors also assume that technologies are born out of conflict, difference, or resistance. Thus most if not all the case studies describe technological controversies, disagreements, or difficulties. The pattern is that the protagonists – entrepreneurs, industrial or consumers, designers, inventors, or professional practitioners – seek to establish or maintain a particular technology or set of technological arrangements, and with this a set of social, scientific, economic, and organisational relations (Bijker and Law, 1992, p.9).

This concept of 'interpretative flexibility', and the social dynamics related to it, seems most important for approaching an investigation of Internet-based urban information systems, given the fact that a 'digital city' is very much a concept – and a technological object – in need of a definition, and that this definition is likely to come during the development and deployment of the cybercity itself.

Technology contributes to its own shaping: the importance of 'paradigms'

Interpretations and visions can also stem from technology itself, or better from the existence of previous or parallel technologies or working practices, that can affect the shaping and the establishment of a new system. As noted by Guthrie and Dutton in their famous study on early development of public information initiatives in the US:

> Existing technology presents a variety of opportunities, problems and constraints that shape the future of technology (MacKenzie and Wajcman, 1985). One way in which existing technology shapes new technology is by providing mental models of solutions. (…) Technological paradigms are the widely accepted, exemplar puzzle solutions, employed as models or examples, which replace explicit rules as a basis for the solution of the remaining puzzles of technology development. (…) The technology designers in our case studies coped with developing these new community information utilities by viewing the projects as analogous to existing exemplar technologies with which they were familiar (Guthrie and Dutton, 1992, p.583).

This yet again reinforces the consideration that technology cannot be looked at as a self-standing fact, but as a development shaped by many different factors. Any piece of urban technology, however apparently revolutionary or innovative, is very probably the product of a process of interpretation and negotiation between competing visions and needs. This involves the presence, and indeed the influence, of previous technologies, early experiments that have become 'exemplar', and accepted practices that can be 'projected' into the new artefact or system.

Changing technology in a changing city

All of these considerations encourage us to do research in order to get knowledge about a phenomenon that too often is seen as a pure fact with few things to understand, other than learning to use computers and networks. It opens the possibility of examining technologies from a wide and multi-disciplinary point of view, going inside cyberspace though not being an engineer or a programmer, and staying outside and yet not trying to look at it as a sure threat. Possibilities open for exploration of design and policy choices and alternatives to them, for a constructive type of criticism that might lead to the formulation of design and policy recommendations. As William Mitchell states,

> Massive and unstoppable changes are under way, but we are not passive subjects powerless to shape our fates. If we understand what is happening and if we can conceive and explore alternative futures, we can find opportunities to intervene, sometimes to resist, to organize, to legislate, to plan and to design (Mitchell, 1995, p.5).

The next chapter will focus on several aspects of the relationship between cyberspace and public urban space. The main topics of the commodification of information, social access to urban cyberspace and the relationship between real and virtual places will be developed, in order to identify key themes to observe the way this new dimension has been, and is being developed. In particular, attention is focused on the World Wide Web, and its early applications in civic environment.

Chapter 3

Exploring the Digital City

Introduction

In the previous chapter we acknowledged the need to avoid a deterministic approach when considering the complex relations between telematics, the city, and traditional urban public space. Consequently, the necessity was identified to focus the research work outlined in this book on the aspects of the social shaping of the 'cyber-city', rather than on the efforts of foreseeing possible future developments of the phenomenon.

Assuming that a digital civic initiative is not to be seen as the product of a neutral technology, as argued in the previous chapter, a series of socio-economic factors contributing to and affecting the shaping of cybercities must be carefully examined. This is one concern of the current chapter, as it highlights the suggestions that come from other researchers as well as trying to find the gaps in the knowledge of the phenomenon of digital cities, mainly due to either lack of empirical research and a too deterministic attitude in writing about virtuality.

This chapter starts from considering the confusion that exists about terms like 'digital city', 'cybercity', 'virtual city' etc. A first step in researching the phenomenon has in fact been working on the definition of a typology of digital cities in order to make sense of various approaches, and not to study different phenomena as if they were homogeneous.

Second, the problems related to ownership and to the presence of private firms digital city development are examined and a parallel between the commodification of public space and of cyberspace is drawn. Third, issues are considered which relate to the differential access to urban cyberspace, and the social polarisation that can derive from different approaches in designing the digital city. In the fourth part the point of view changes, from the aspects of the shaping of urban cyberspace as mere electronic territory, to the relation that it can develop with its physical counterpart. The interactions between physical and virtual space are considered from the point of view of the 'defragmentation', or integration of the city either as a group of communities and as a piece of built environment.

All of these issues help in trying to identify the key questions that had to be investigated and answered in order to get an insight about how digital cities were being shaped at the early stages of the development of urban cyberspaces.

Finally, closing the circle, considerations are made on a possible way to approach the construction of the much-needed typology of digital cities in the light of the topics and issues faced in the chapter. Relevant criteria for the classification of initiatives, as well as important issues that have been leading the rest of the study, are highlighted.

Towards a typology of cybercities

As already mentioned, cyberspace has been seen by many as a potential new type of public space, in which interaction among people can be set free from the problems generated by physical space and by the contemporary fragmented city. Several kinds of so-called 'virtual communities' have been developing, as more and more people are getting connected to the Internet, beginning to share a virtual space that allows them to communicate without the need to meet in some physical place. 'Perhaps people explore these networks for the conviviality that they find hard to discover in contemporary cities because of the shift towards commercialisation, the "packaging" and "theming" of urban landscapes, and the wider "fortressing" of affluent neighbourhoods?' (Graham and Marvin, 1996, p.230). This search for a safe and enjoyable environment, together with all the hype about the positive effects of cyberspace on cities that could found daily on the different media, especially towards the end of the 1990s, have made the spatial and above all urban metaphor, a dominant characteristic of the early development of many virtual sites present on the Internet.

It has become a sort of fashion, or even convention, to associate certain sites on the World Wide Web with the idea of a precise public place or territory, such as a square, an avenue or even an entire city. Consequently, a whole generation of self-claiming 'digital cities', 'cybercities', 'cyberdistricts', 'virtual squares' and 'electronic agoras' has in the past few years flourished on the net. Indeed, the spatial metaphor tends to work so well, giving an identity to electronic services, that speaking about a World Wide Web 'site' has become natural. No surprise then that Time Warner decided to call their interactive chat software for the Internet 'The Palace', and chose to provide people the ability to create their own virtual palaces, containing 'rooms' and 'objects' to interact with, that could be visited by several guests at the same time and that can actually became virtual gathering sites.

Figure 3.1 Screenshot of 'The Palace' Internet chat software

Having downloaded the software from the Internet and registered, at the cost of 20 dollars, every user was able to create his/her own 'world', deciding the shape and the rules that visitors will have to deal with. This piece of software, born in 1995, was not successful as a commercial enterprise, and Time Warner eventually gave up the subscriptions scheme. However, it became rather successful as a free software tool for creating virtual chat rooms, and is still in existence at the moment of finishing this work. It can be downloaded from www.thepalace.com.

It is hard to be a digital city

However, visiting a 'Palace' cannot mean going into a public space just because it allows some type of communication among different people. Rather, it is much more similar to visiting a private house – hence its very appropriate name 'palace' – in which the landlord is definitely in control of everything. Although it is a very smart and possibly useful piece of software, in a way it represents the mirroring within cyberspace of post-modern fragmentation over public, democratic places. Do you want to meet me? Come to my house and behave yourself...

Another interesting case was the Barclaysquare, a WWW site launched in May 1995 using the metaphor of the square to present what was actually a virtual shopping mall and, basically, a good looking graphical menu leading to a series of shops and facilities on the Internet.

Figure 3.2   **'Barclaysquare', an early experiment of commercial virtual environment**

Barclaysquare was successively modified and eventually abandoned, but other similar examples have been developed using the marketsquare metaphor, to the point of developing sophisticated three-dimensional, fully navigable virtual shopping malls, through VRML (Virtual Reality Modelling Language) technology, such as the 'Envisage' Shopping Mall.

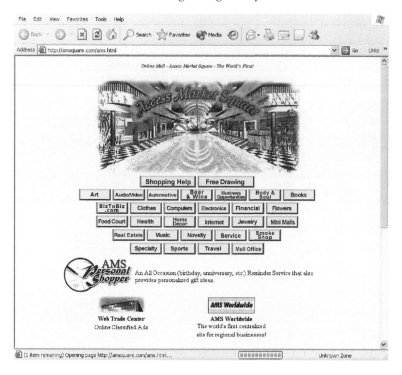

Figure 3.3   'Marketsquare', another example of urban metaphors used for commercial sites

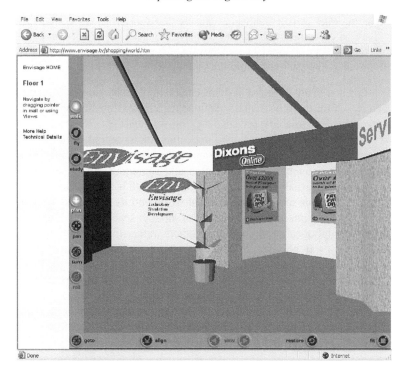

Figure 3.4   The 'Envisage' virtual mall, based on three-dimensional VRML technology

It is hard to say how Barclaysquare could be referred to as a square. Was it possible to meet people there? To play saxophone begging for some money? To shout your ideas about politics or society? To know other people's ideas? To fall in love? Obviously nothing like this was feasible in such a virtual, commercially-oriented place. Your credit card number was your only personal characteristic that could be considered relevant and really valuable in such a 'piazza', and you would be able only to meet goods on sale, or rather the descriptions of them.

These two examples, the first very 'personal' and smart, with a possibly good level of interactivity, the second much less interactive and controlled by a big private company, have something in common. They both showed that speaking about 'squares' and 'streets', 'palaces' and 'halls', 'cities' and 'districts' could be in cyberspace even more ambiguous than in real physical space, and that it did not imply that certain labels corresponded to genuinely 'public' sites. And sometimes even considering those telematic services as 'places' is really hard and may look inappropriate, since they are really only glamorous menus, as can be understood by this description made by Michael Batty of a certain 'cybercity':

In short, Cyberville is an intelligently organized entry point to the web for users who wish to focus on the sort of services one can access electronically. For example, if you click on the stock market, you can eventually find yourself at M.I.T.'s AI Lab where they have an experimental stock market analyzer, indeed of you click on any of these items, you can begin to travel the web moving from site to site according to the structure embedded into each homepage, but directed initially by Cyberville (Batty, 1995).

It is a fact that many different systems can be implied by names like 'cyberville' or 'cybercity' . Some can be actually defined as 'cities as menus', useful to access several worldwide-available services, and borrowing the spatial metaphor just because it works quite well and it is appealing for potential customers/visitors. Others can be configured as stand-alone virtual spaces, completely displaced, having potential for human communication and sometimes being even organized like a city, with a 'virtual' council and mayor, but with a very weak relationship – or even no relationship at all – with the real world and its problems. They actually look more like a game or a new type of hobby space, with a potential for socialisation, mainly geared at computer addicts, rather than something that could replace or in any way have an influence on real towns. The case of 'Cybertown', for instance, is paradigmatic, as this initiative has been a very successful and evolving entertainment site with a very effective use of the urban metaphor. Cybertown (www.cybertown.com) has been providing for a rather long time an Internet 'place' for playing games, establishing a 'virtual' presence and identiy – including a three-dimensional virtual apartment – shopping, socialising and even having a virtual pet.

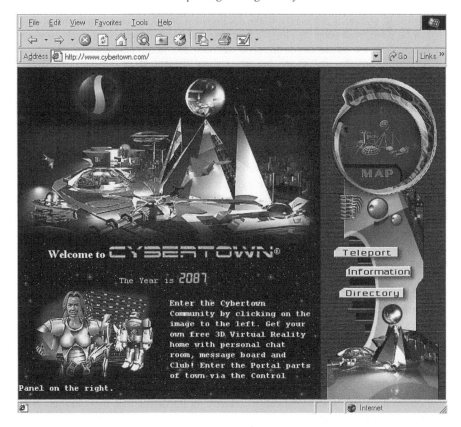

Figure 3.5 **'Cybertown', a complex, interactive virtual environment with no links with any physical space**

Cybertown is nowhere, apart of course from the fact that its software must reside in a server located somewhere, but this is irrelevant to its 'inhabitants'. It is of course virtually everywhere, as it can be accessed by any Internet terminal, but surely being 'everywhere' means that it has no specific relationship with any physical place, its culture, or the people who live there.

Further, there are digital cities that constitute an attempt made by real city councils or private firms, to regenerate communication among citizens and between citizens and the administration, to give new life and visibility to the different local communities, to give visibility to the whole town itself in order to be able to compete at a national or even international level. These two very different approaches can be defined as the building of 'non-grounded' and 'grounded' digital cities:

> First, non-grounded web cities use the familiar interface of a 'city' as a metaphor to group together wide ranges of Internet services located across the world. Second, 'grounded' virtual cities are actually developed by local agencies to feed back

positively, and relate coherently, to the development of specific cities. Such virtual cities can be configured either as glossy advertising and promotional spaces, with no useful information for residents, or as civic services providing public electronic spaces supporting political and social discourses about the city itself (Graham and Aurigi, 1997).

Digital cities of the latter type are the object of this research, as they are seen by the councils and agencies that are promoting and implementing them as potential instruments to improve life in the physical town. Nevertheless, the presence of sometimes radically different types of virtual cities, and the frequent abuse of such a definition, creates the need for a rather precise categorisation of cybercities, in order to spot those that actually have a direct relationship with public urban spaces, getting rid of the confusion and the approximation coming from the uncontrolled hype offered daily by the media.

As a result, a first research theme arises here. How can we develop a typology of digital cities, and what are the main differences among them? And what are their relationships with the built environment of cities?

The commodified space

The case of Barclaysquare, and the many others similar to it, suggest the need to think about one of the most important processes to consider when observing how virtual sites are being shaped: the transformation of public space into a commodity, into something that can be sold or that can be an instrument to sell things, a phenomenon called 'commodification'. There is no doubt that the Internet, born years ago as a military and academic network, has become a mass phenomenon and that many private firms have been getting more and more interested in it. 'Nearly 40% of the networks connected are in the commercial domain, with less than 30% in the educational, illustrating the economic importance of the net and the intense debate currently concerning its ownership, funding and possible privatisation' (Batty and Barr, 1994, p.703) wrote Batty and Barr on *Futures*, and the figures have kept continuously increasing.

The World Wide Web in particular, the graphical and hypertextual sub-system of the Internet, has been increasingly used as a powerful tool for trading at all levels, and it is a commodity in itself. A whole market developed quickly from the late 1990s onwards, around the offer of network spaces, web sites design, development of increasingly secure software for transactions over the Internet. Reports like 'Brands on the web', made by Forrester Research, a Massachusetts-based consultancy, predicted that 'by 1997 three-quarters of the top 200 brands in the United States "will promote themselves online". A quarter are doing so already' (Azhar, 1995). This trend reinforced itself in the following years, with the Internet becoming a crowded marketplace where people would spend their real money and buy goods, rather than just being targeted by electronic advertising and promotion:

The future of e-commerce continues to brighten, according to Jupiter's research. Driven by online population growth, increased spending by buyers, and higher percentages of online shoppers (142 million consumers or 65 percent of the online population will have made a purchase online by 2007) Jupiter expects online retail spending to reach $105 billion by 2007, accounting for 5 percent of all U.S. retail spending and influencing 34 percent of all U.S. retail spending (Greenspan, 2003).

Even the Internet itself has been going towards privatization, as envisaged by Batty and Barr. A big milestone for this can be identified in what happened in the United States in April 1995. The National Science Foundation, a government-funded organization that used to manage the so-called 'Internet backbone', the American 'gate' of the net, handed over the managerial reins to three commercial carriers: Sprint, Ameritech and Pacific Bell, and it was noted that although 'the changeover should be transparent for the users... its symbolic significance is immense' (Cottrill, 1995).

Not 'Big Brother'

All of this clearly demonstrates that the Internet, and particularly the World Wide Web, has been more and more influenced by the private sector. This does not mean necessarily that the net should be seen as the actualisation of the Orwellian 'Big Brother', a centralised system able to heavily control our lives. Nick Land claimed that

> There is a very similar pattern that you find in the structure of societies, in the structures of companies and in the structure of computers, and all three are moving in the same direction. That is, away from a top-down structure of a central command system giving the system instructions about how to behave, towards a systems that is flat, which is a web, in which change moves from the bottom up (Land, from *Visions of Heaven and Hell*, Channel 4 Television, 1994).

But, apart from these very optimistic conclusions, does this imply that a space based on, or enhanced by, telematics is going to be democratic and truly public? The absence of a central control room might even simply highlight a high level of fragmentation, in which specific 'fragments' end up being privileged respect to others. As the postmodern city is characterised by disaggregation and influence of the private sector on specific, appealing and profitable bits of public space, this can be true for cyberspace, as well. Schiller for instance noted that

> Carrying over into cyberspace as sponsors flexed their muscles, in turn, are practices that have long since become customary throughout conventional media. Who said advertisers have an obligation to support all the publishers that may choose to throw in with them? Advertisers clustered around the most heavily trafficked web portals, where they could reach the largest needed audiences; by one account, these leading sites raked in nearly three-fifths of Internet ad dollars by 1998. The effect was, predictably, to put financial pressure on less popular sites, much as advertisers' preference for the cost

efficiency of a municipality's leading newspaper once helped to transform newspapers into local monopolies (Schiller, 1999, p.133).

The problem with commodification, both in physical and virtual space, is not trying to avoid a sort of 'big brother effect'. It is more likely to be dealing with the co-existence of public and private in the same spaces, and in the balance between the often diverging interests of the two sectors. It is realising that the system cannot be 'flat', as Land tried to describe it, and that economics count in determining geography and the relevance of place, including cyber-places. It can also be dealing with the processes that allow a square hosting homeless people as well as drug dealers, to becoming a shopping mall or a sports centre with no pushers but no poor people at the same time. In the case of Berkeley's People's Park described by Don Mitchell, the local university decided to create volleyball courts taking control of the space:

> Activists and the homeless people who used the Park promoted a vision of a space marked by free interaction and the absence of coercion by powerful institutions ... The vision of representatives of the University (not to mention planners in many cities) was quite different. Theirs was one of open space for recreation and entertainment, subject to usage by an appropriate public that is allowed in (D. Mitchell, 1995, p.115).

There is also the strange situation that is created by those sites that tend to be hybrid, being referred to as semi-public or semi-private, that generate, eventually, a loss of space for whole categories of disadvantaged citizens.

> In the course of the 1980s the distinction between public and private space became less and less clear, with the multiplying of 'private public spaces': that is, privately owned and managed spaces offered for public use. (...) In fact is probably more appropriate to interpret the 'private public space' phenomenon as a sign of a tendency towards the privatisation of public space (Bianchini, 1988, p.5).

Although this quotation by Franco Bianchini dealt with privatization processes going on in actual cities in Britain, it is easy to claim that problems of the same kind could affect virtual space, and that we should be concerned about them, while observing how cybercities are being built up.

Promotional spaces

The more we observe virtual worlds the more it seems possible to draw a parallel with some aspects of the physical ones, rather than presenting them as a revolutionary replacement of the latter. For example, Christine Boyer speaks about public spaces strongly influenced by the private sector, and often shaped by it:

> The city these spaces represent is filled with a magical and exciting allure, landscapes of pleasure intentionally separated from the city's more prosaic or threatening mean streets. Controlled by the rules and values of the market system, these places offer a diet of synthetic charm that undermines critical evaluation. ... As old-style 'public space'

declines and popular control of the streets becomes a thing of the past, a new-style 'publicity' or 'promotional space' evolves on which the reputation of the sponsoring corporation is visualized and its production of 'civic values' promoted (Boyer, 1993, p.119).

This raises the question of whether similar types of urban design are actually taking over cyberspace. Electronic space suits these publicity needs; it is potentially a very effective 'promotional space', as demonstrated by the higher and higher number of even small private firms interested in advertising themselves on the World Wide Web. Therefore, virtual sites could just be something very similar to those shopping malls that '...were characterised by a tendency to confine public social life to certain locations, certain hours and certain categories of "acceptable" activities' (Bianchini, 1988). This aspect of the debate has been particularly alive in the US, where 'We find mall aspects in theme parks, themed environments. These provide safe, secure environments where people can interact. It looks very much like public life, but in fact it really isn't, because the environments are owned and controlled and heavily regulated by, generally, very large global corporations' (Dewey, from *Once upon a Time in Cyberville*, Channel 4 Television, 1994). And cyberspace has been seen as the ideal evolution of this kind of themed and heavily controlled spaces as Fred Dewey again states:

> People interact somewhat randomly, but the actual experience is entirely manufactured – all of its terms are defined ahead of time, the experience is very similar to that of going through virtual reality. While this provides a kind of vitality, at the same time it's based on leaving behind the mess of real urban life. Everyone expects, somehow, that the Cyberworld is not going to have these kinds of parameters and controls. This is extremely unrealistic, I think (Dewey, from *Once upon a Time in Cyberville*, Channel 4 Television, 1994).

Merging public and private

If this is a fact to be considered, and the worries about an only apparently public space can be shared, it is true as well that the commodification, or should we say privatisation, of spaces, supposedly public, is not just a recent phenomenon. The Greek agora or the Roman forum, often taken as examples of our lost capacity to create democratic spaces, were themselves places of exclusion and control, as Don Mitchell remarked, and in a way they were commodified, in the broad sense of 'controlled by the interests of an elite'.

> The public that met in these spaces was carefully selected and homogeneous in composition. It consisted of those with power, standing, and respectability (...) In Greek democracy, for example, citizenship was a right that was awarded to free, non-foreign men and denied to slaves, women and foreigners. The latter had no standing in the public spaces of Greek cities; they were not included in 'the public' (D. Mitchell, 1995, p.116).

The city of Florence in the fifteenth century, to use another example, was in itself a very controlled city, and the magnificent palaces we are able to visit today were private residences or offices that could not simply be accessed by everyone. Consequently, the beauty of the public places of Florence is due to a composition of private elements owned by the mighty families who used to rule the town, or semi-public spaces like churches and chapels, or the streets themselves, that were characterised – and still are – by rows of shops, banks and workshops. These were owned by a powerful class of traders and craftsmen, able to have a strong influence over the decisions concerning public space in the city centre, during the Renaissance period as well as at present times.

It is extremely important, therefore, to get rid of the excessive idealisation about public urban space, as it cannot be seen as a pure realm untouched by private interests, unless we commit ourselves to rethink completely the basis of Western civilisation. On the other hand, we must not underestimate the legitimate worries about the current situation, in which the quasi-public is shifting towards the completely private, and many cities are being increasingly characterised by fortified spaces, while the open ones are being abandoned. It is in a way a matter of shades, of control and equilibrium between the economic boost coming from the private sector and the need for public, accessible, local culture friendly space. Policies have to be thought from this perspective of a reality in which public and private merge and interact generating a very complex situation, as it has been claimed by Elizabeth Wilson:

> ...urban industrial society has always provided spaces that are neither public nor private, yet both: classically the café, the museum or concert hall, the department store. The airport and the mall are latterday equivalents. We therefore need a more nuanced way of discussing space than the dichotomized public/private conceptualization... (Wilson, 1995, p.159).

This overlapping of public and private is what determines the actual shaping of places and it is to be considered when it comes to observing cyberspace and digital cities, especially if the virtual sites we are interested in are those with a strong link with real towns, as are civic networks and councils' web sites. Despite the hype, cyberspace is not a pure realm. Thus a digital city is likely to be shaped by both private and public interest, and by the interaction between the two operated through policies and market dynamics. Citizens surfing the Internet will never find themselves in an ideal, market-free world. To engage with the Internet they have had to buy the necessary equipment on the market, very possibly to subscribe to a private Internet provider, and eventually access the highly 'promotional' spaces we were have been talking about. Consequently, Internet dwellers choose, with every click of their mouse, between public and private elements, pages, interconnected virtual spaces. The problem with virtual urban space could be seen as being able to create 'squares' with availability of shops and civic rights at the same moment, goods and information, sales and free human interaction rather than completely private and controlled malls in which the only option is buying something, whatever it is. The ideal, trade-free, interests-free square would probably turn into an empty place, as can also happen in the physical city.

The dynamics of this process need observation, as well as the policy choices that can affect them. The risk of the private sector having too strong an influence in the shaping of public cyberspace was well acknowledged in the TV program *Visions of Heaven and Hell*, broadcast by Channel 4 Television in 1994:

> The technology of the future may be more user-friendly, but it will still be supplied to us by companies that have their own commercial interests and alliances. What is best for them and their shareholders may not be best for us (*Visions of Heaven and Hell*, Channel 4 Television, 1994).

There is no doubt that private money can help local administrations in regenerating parts of the city, but some choices can be questionable. Several needs can be met, but many others can be definitely ignored depending on the relative interests and on the advertising capabilities of certain places, as suggested by Christine Boyer:

> If private capital is able to subsidize and sponsor the public arts program in Battery Park City, for example, where sculpture gardens, horticultural displays and decorative lighting have been designed, then we must also be aware that since the late 1980s public parks throughout the city are littered with broken glass, trash and abandoned cars, while the employment of maintenance workers has been reduced dramatically and the expenditure of tens of millions of dollars delayed (Boyer, 1993, p.117).

Could this happen in cyberspace, as well? A poorly regulated private intervention in the shaping of public sites could result in the improvement of only those aspects of the service that have an appeal for trading, selling, promoting goods or anyway for the business sector, possibly leaving behind non-lucrative social needs and features. It is not by chance that a certain utopian and blindly enthusiastic thinking about virtual space is based on a strictly capitalist vision of a self regulated market, in which the people's needs are identified with products to buy and with the values of the society of consumption:

> We are dictating what products we want, and they have to follow suit. (...) The people, the CEOs running the companies, are not the dominators of culture. We, the individuals, are the dominators of culture, and they are our servants. They are slaves to the dollar (Rushkoff, from *Once upon a Time in Cyberville*, Channel 4 Television, 1994).

According to Rushkoff, being the dominators of culture means being able to influence the market by purchasing goods. This implies also that those who are not able to buy are to be considered out of the social dynamics, completely impotent in front of the market based society. According to such a vision, people ratify their existence as well as some form of visibility at the very moment they can act as consumers. And also among those who can spend money, the ability to 'dictate' how cyberspace should develop seems to be proportional to how wealthy they are. Rather than denying the commodification of virtual space, as it tries to do, this

point of view confirms some possible basic inequalities when it comes to dealing
with the net as a deregulated marketplace, as clearly stated by Elizabeth Wilson:

> Economic and social inequalities remain as gross as ever, yet the global shopping mall
> renders them curiously invisible. Those without the passport of money are simply in
> absence... Invisibility is a crucial feature of modern inequality (Wilson, 1995, p.158).

All of this suggests that the phenomenon of the commodification of public areas
can affect cyberspace-based sites and services even more than the corresponding
physical places, because of the intrinsic 'promotional' potential of virtual space and
of the interest of the private sector in investing in cyber-activities. Signs that the
ICT industry is inevitably very keen on 'cherry-picking' customers and offering
faster and better service to those who are endowed with disposable income, are not
difficult to notice. Nevertheless, as suggested by Wilson, this is partly inevitable
and does not necessarily spoil the social potential of cyberspace in creating virtual
public spaces, bearing in mind that the 'pure' public as the absence of any kind of
private influence is actually just a concept that does not find a correspondence in
the real world.

A question therefore arises when considering how digital cities have been
shaped in the early stage of their development, about the relationship between
public and private sectors. How have these sites been shaped in terms of the
merging between common and particular interests? How have the different public
administrations been trying to regulate the presence of private actors in the process
of shaping and managing public cyberspace? And what has been the influence, if
any, of the private actors in the design and management processes of these sites?

Competing cities

The phenomenon of the commodification of cyberspace has not just been driven by
the action of private firms that advertise themselves on the net, or use it to sell
services and products. Although the competition among traders and corporate
societies is on computer networks, a similar phenomenon has been occurring in
cities themselves, especially in the Western world, although not exclusively.

Problems like safety and pollution, together with dramatic changes in the
economy and labour organization, are inverting the growth process of many
contemporary cities. Dilys M. Hill got to the point of claiming that:

> In the electronic age the centuries-old rationale for cities is vanishing. People and
> employers find it cheaper to move work elsewhere, and residents depart. If nothing is
> done to make cities safer and more attractive, their decline will mean an increasingly
> polarised population of rich and poor, with a deterioration of public services and
> increased racial tension (Hill, 1994, p.243).

While Hill acknowledged some crucial social problems that can be generated by
this exodus of wealth from cities, this statement seems too extreme and hardly
matching what is happening in reality, as cities are possibly more than ever on the

agenda, and keep growing rather than vanishing. It seems credible, however, that rather than witnessing a scenario of vanishing cities, we have to redefine geographies and the dominance of determinate centres on the basis of their capacity to regenerate themselves, as well as exploiting the potential of new technologies.

Still questioning the relevance of the city as a place facilitating human proximity, some researchers tended to see the ability of being 'displaced' and moving anywhere, by being 'wired' to the net, as a specific, positive aspect of freedom for the new century:

> People can live almost anywhere they wanted without forgoing opportunities of association or useful and fulfilling employment... important social needs would be all available 'on-line' without waiting 'in-line', whenever and wherever you need them (Ogden, 1994, p.723).

But 'almost anywhere' actually means that it is possible to move only to those places, or those cities, that are able to offer certain facilities and characteristics. At the same time it means that sites with poorer features than others are cut off, and that certain areas can suffer even more from being 'technologically excluded'. Economies of scale, however, suggest that cities that gather many people are much more likely than remote rural locations to be endowed with high technologies at cheap-ish prices. Cities can also offer at the same time the considerable advantage of physical proximity, and the ever-important face-to-face interaction that this can allow, together with electronic communication potential. Consequently, cities – above all – find themselves in a big competition, often transcending national borders, to acquire people and wealth, as was clearly stated in the brochure advertising the 1988 Remaking Cities Conference, organized jointly by the American Institute of Architects and the Royal Institute of British Architects:

> Businesses and individuals – increasingly free to locate where and when they want – select cities with the finest features and benefits. They look for history, culture, safe neighbourhoods, good housing, shops and education, and progressive local government. Cities are competing, and their edge is livability (quoted in Boyer, 1993, p.125).

What are the main ingredients of 'livability', that can attract people to one city rather than to another? Several of them have been listed in the previous quotation, but as mentioned before a new variable is getting more and more relevant: communication. In the post-industrial era or, as many have begun to call it the 'information age', the ability to get wired and stay in the flow of the worldwide information is a crucial feature. It allows towns to retain or attract more and more firms, some of them being big corporate giants looking for more convenient places where to establish their regional, national or even continental headquarters. This is why: 'many cities... are trying to gain a competitive advantage by getting themselves wired with fibre optic cables before their rivals do' (Batty and Barr, 1994, p.701).

It is not only the big firms which are interested in the technological facilities offered by a city, in terms of hardware capabilities and bandwidth, in order to keep their contacts with the rest of the world. Many medium and small sized enterprises can benefit from 'city software', too. Civic and regional networks, cybercities and cyber-counties, can in fact provide vital bits of information for local businesses. Facilities to allow remote communication with banks, the public sector, customers and other firms have been seen as a dramatic feature of contemporary urban and metropolitan areas. Together with these practical advantages, the ability of local agencies to endow their place with a forward-looking images and a feeling of future-proofness, has become a very necessary feature within this competitive climate.

Providing civic networks and World Wide Web digital cities is a chance that has been considered by an great number of councils and regions, in the US as well as in Europe. And for the reasons stated so far many, if not all, virtual towns have been aiming at providing business opportunities, as demonstrated by the case of Nottingham:

> Nottingham claims Emnet is the country's first regional commercial computer network. Unlike earlier rivals, such as Manchester's Host, it is aimed at the business community. Developed by Nexor, a computer consultancy, it uses the World-Wide Web for trading and marketing. For companies like Boots and NatWest, who have their own pages, it provides a vehicle for home shopping or banking. Emnet might also become a development agency, advising on telecommunications technology (Hanna, 1996).

But these business-oriented digital cities could become places that are commodified indirectly. Although they have been designed by the public sector, their aims can be other than benefiting the city as a whole and relieving it from fragmentation. Big companies like Boots or NatWest are not always necessarily interested in local social problems and, as the primary aim of the system is business, it is likely to be tailored to their needs, rather than to those of the communities living in the area.

Can there be at least an advantage in terms of jobs to local people? Perhaps, but it is fair enough to be be questioning this as well. Speaking about investments in American towns, Christine Boyer, for instance, warned about the dangers of the 'global' competition game to lure big corporate investment in, noting that:

> ...these super corporations have neither loyalty to a specific nation nor social accountability to any one locality. Many of these transnational corporations are major investors in the redevelopment of old American downtowns. Investment today, however, may bring abandonment tomorrow (Boyer, 1993, p.115).

From this, a distinction should be drawn between cities that have been implementing their business-oriented web pages hosting mostly big corporate companies and those that have been actually seeking synergies with medium and small local firms, retailers and enterprises. The latter have stronger links with the city and its citizens, they are active part of the local economy that can actually bring permanent benefits to the community.

The choice to design a civic or regional information system, whether it is based on the web or on another technological platform, in which the main purpose is providing information to and from business actors, can lead to the final choice of the commercialisation of the information offered. Although making businesses pay to retrieve data and advertise themselves on a public network can seem a perfectly sensible choice, the commodification caused by competition between cities can go as far as seeing the web 'public' sites as a good way to make money, independently by who is using the service:

> Governments are a major provider of free information, and there are fears that commercial pressures, coupled with Congress's fervour for cost-cutting, will force official agencies to charge for public information. Already, some companies have teamed up with cash-starved local authorities in the U.S., with a view to charging for official information (Cottrill, 1995).

This tendency, for some public administrations, to commodify the information offered on information technology systems, together with those design choices giving clear priority to businesses rather than to citizens, can be crucial in the shaping of cybercities. It can influence their aims and their effectiveness, reversing perhaps the expected results of increased social cohesion and better communication between the public and administrators. This leads to two key questions to address.

How have municipalities and policy-makers been relating their digital cities to their local economies? Have they been aiming to revitalise local enterprises and local economic actors or have they been just selling cyber-places to big corporate companies, based elsewhere? What policy processes have been used to engage with local small firms and large multi-site firms?

How have designers and managers of digital cities been addressing the tension between the need to earn revenue from the information that is made available and the need not to restrict social access?

If it is public, I can go there: access to cyberspace

All the problems related to the commodification of public cyberspace lead us to make a further step in this analysis about the shaping of digital cities: the consideration of issues about access and social polarisation. This can be actually regarded as a central problem in the implementation of civic networks and cybercities, above all because of their supposed ability to re-establish or anyway improve communication among citizens and between citizens and public administration.

In this effort to create a new type of common space, considering issues about property and private influence is a necessary but not sufficient step. As William Mitchell stated:

> ...urban public space is not merely un-private. (...) A space is genuinely public, as Kevin Lynch once pointed out, only to the extent that it really is openly accessible and

welcoming to members of the community that it serves. It must also allow users considerable freedom of assembly and action. And there must be some kind of public control of its use and its transformation over time. The same goes for public cyberspace, so creators and maintainers... must consider who gets in and who gets excluded, what can and cannot be done there, whose norms are enforced, and who exerts control (Mitchell, 1995, p.125).

So, to invert the current trend towards social and spatial fragmentation, public space must offer a potential for gathering people while avoiding discrimination. But space, like technology, is never neutral. Through its physical organization it can present itself as unfriendly for certain disadvantaged categories like, for instance, elderly people, children or disabled. Is the cyberworld so different a type of space to automatically set itself free from these limits?

Much concern has been expressed about the threats to democracy in cyberspace. Ogden stressed that: 'More people must learn about the intellectual, social, commercial and political leverage presented by participation in the communications revolution that is cyberspace – while the freedom to do so still exists' (Ogden, 1994, p.714). Similarly, one of the enthusiastic 'gurus' of cyberspace, Howard Rheingold, had some doubts about the future:

> But telecommunications give certain people access to means of influencing certain other people's thoughts and perceptions, and that access – who has it and who doesn't have it – is intimately connected with political power. The prospect of the technical capabilities of a near-ubiquitous high-bandwidth net in the hands of a small number of commercial interests has dire political implications. Whoever gains the political edge on this technology will be able to use the technology to consolidate power (Rheingold, 1994, p.278).

This point of view, although arguably worth considering, embeds the weakness of considering technology neutral or even positive in itself. In this view cyberspace has been seen as a pure, absolutely good development, that might become eventually corrupted and misused. Electronic democracy could turn into electronic dictatorship by the action of some 'big brother'. Otherwise, the net has been and is a wonderful place, offering plenty of opportunities, and the implications how it has been shaped are not very much considered. Neither is the social context in which cyberspace has been developing. That is probably why Ogden himself claimed emphatically: 'It's here and it's now and it's no longer just for computer scientists; it's computing for the masses. Welcome to cyberspace!' (Ogden, 1994, p.714).

'It costs too much'

But what can the masses do? Before welcoming people to a virtual world, it is fair enough to determine who can actually afford to access that space. Observing the crisis of the 'physical' public space in Britain, Franco Bianchini noted that:

> A study carried out in 1987 in a large Sheffield council estate, with male unemployment and car ownership rates of 30.4% and 15% respectively, found that '(it) costs too much'

was by far the main reason given by residents for what prevented them from going out (Bianchini, 1988, p.5).

Why should cyberspace not obey the same rules? If those disadvantaged citizens are discouraged from going downtown because of their very low budget, could they not find problematic accessing the 'electronic downtown' for the same reasons?

Being able to 'surf the net' from home implies a series of expenses. You need to buy a personal computer and some type of modem, subscribe to an Internet provider for access to either narrow or broadband, and in most cases own a telephone line. If a municipality chose to charge citizens for the use of its system, as was envisaged in the previous section about the commodification of cyberspace, that could possibly be a heavy additional cost. But there is evidence that many families living in deprived urban areas have been having problems even to afford to pay for the first and most basic facility, the telephone line:

> In Cruddas Park, a Newcastle upon Tyne council estate of some 3,000 households, it has been estimated that the connection rate is just 26 per cent. As recently as last year, the government still considered non-connection to be primarily a rural problem, but the geographical and social extent of the unphoned phenomenon is now, quite clearly, a key problem in our marginalised urban areas (Dyer, 1995).

How, then, can be these families be expected to fulfil the hyperbolic prophecies of Timothy Leary when he stated:

> The social and political implications of this democratization of the screen are enormous. In the past, friendship and intimate exchange have been limited to local geography or occasional visits. Now you can play electronic tennis with a pro in Tokyo, interact with a classroom in Paris, cyberflirt with cute guys in any four cities of your choice. A global fast-feedback language of icons and memes, facilitated by instant translation devices, will smoothly eliminate the barriers of language that have been responsible for most of the war and conflict of the last centuries (Leary, 1994, p.76).

The 'new breed' Leary was talking about are still included in a privileged minority that can, first of all, afford all the necessary expenses to spend perhaps hours every day playing tennis with a Japanese or attending a virtual class somewhere else. Sheridan M. Tatsuno goes further, envisaging a

> ...multimedia city, a global networked city in which information is pushed down the social pyramid to the poorest person. It is not just a high-tech media city like Hollywood or Silicon Valley, but also a low-tech, high-touch city like Bangalore, New York's Harlem, or Lima's barrios (Tatsuno, 1994, p.201).

But just after this he clearly shows the limits of his utopia saying that: 'Indeed, the ultimate multimedia city is not a place, but a state of mind. It is a city in which information is easily accessible to the average person' (Tatsuno, 1994, p.201).

And it is certainly hard to recognize those who live in the barrios as 'the average persons', as well as the concept of creating a useful cybercity not as a place, but as a 'state of mind'. Could this change? The idea that technological progress makes computers and electronic appliances cost less and less, becoming affordable for everyone, is one of the main arguments underpinning the hype about cyberspace and cyber-democracy. But, although computer prices seem to dicrease constantly, it must be said that to run up-to-date software, more and more computing power is required. The Internet itself evolves continuously. If it was possible – when it all began – to download information using a 80286-based PC and a low-spec modem that would now cost literally a few pounds, accessing state-of-the-art sites requires much higher computing power, sophisticated graphics cards, and above all a lot of communication speed, such as a broadband connection. This is also due to the ever-increasing sophistication of the World Wide Web, that has evolved from a relatively simple environment made of static text and images, to a truly dynamic, multimedia domain. Simpler sites have quickly become obsolete for most of the information present on the net, as firms and institutions publish in increasing sophisticated ways, literally forcing users to upgrade both hardware and software. The result of all of this is that the costs to be 'up and running' are not actually changing that much, but that older and cheaper devices are actually set aside and made obsolete by the market-driven technological evolution. As Julian Stallabrass notes:

> There is no interest in selling electronic commodities at the price the world's poor can afford, nor is there likely to be. The idea that high-band global networking will become truly universal in a world where only a fifth of the population currently have even telephones is laughable (Stallabrass, 1995, p.11).

This is not true just for the third-world countries. We have already seen that deprived citizens in Britain could have similar problems to access cyberspace. More generally, the idea of a free market offering cheap telematics for everybody is flawed even in this wealthy part of the world. For example, research carried on by Inteco in 1995 in five European countries, including Britain, indicated that: '...the home PC boom, on which most manufacturers are counting, is thus illusory. This year in the UK, about 77 per cent of sales will go to householders upgrading their existing PCs or buying additional ones', and that, concerning cyberspace: 'It found that only 10 per cent of Internet access was from home, indicating that "surfing the net" is overwhelmingly done on other people's phone bills' (Bannister, 1995). Although since the mid-nineties new commercial offers, such as 'flat' or unmetered Internet tariffs have appeared on the market, facilitating at least some to accessing cyberspace more and more often, it can still be argued that universal access is far from being achieved, and an elite of privileged people has been using the net and, consequently, has influenced its shaping. The first, necessary requisite to dwell in cyberspace is wealth, and this has been a crucial problem addressed by those organisations determined to create civic networks able to involve as many citizens as possible. This leads to a further question.

How, then, have policy makers been dealing with the access problems to their civic cyberspace, for those who are economically disadvantaged? Through what policy instruments have the socially disadvantaged groups been given access to digital cities?

Being visible

But although having access to the Internet and to the information that is contained within it, is a primary condition to be a cybercitizen, it is probably not enough. 'Access' in fact does not mean just 'being able to go somewhere'. Especially for public space, it implies the possibility to use that place, being able to express yourself without, of course, harming others. If we go back for a moment to the definition given by William Mitchell who said that: 'It must also allow users considerable freedom of assembly and action' (Mitchell, 1995, p.125) it becomes clear that, beyond the first step of 'getting there', the ability to be an active, involved participant of cyber-life plays a very important role.

And this must have been the feeling of Howard Rheingold when he first approached the virtual community called The Well:

> ...I soon discovered that I was audience, performer, and scriptwriter, along with my companions, in an ongoing improvisation. A full-scale subculture was growing on the other side of my telephone jack, and they invited me to create something new (Rheingold 1994, p.2).

This positive personal experience made Rheingold himself provide a wide generalisation when he stated in a TV interview that: 'When you plug that computer into that telephone, you've not just created a printing press, but a gathering place, a place of assembly, a place where people can exercise free speech and debate with each other' (Rheingold, 1994, from *Visions of Heaven and Hell*, Channel 4 Television).

But talking about 'that computer into that telephone' brings with it a great deal of over-simplification. Provided, as was said previously, that someone could afford to have both computer and telephone line, and maybe a broadband connection, would he or she be educated enough to use them properly? As Graham and Marvin noted: 'Network traffic is also often dominated by the technologically "converted": a very small and extremely unrepresentative set of people' (Graham, Marvin, 1996, p.232). This means that culture and education, and more specifically computer literacy, can have a substantial influence over the ability to become an active dweller of cyberspace. Therefore, the access related issues are not limited to those of direct economic nature. Access involves problems inherent to the whole of the social sphere and dealing with it requires the adoption of a more sophisticated viewpoint, rather than just considering those who can 'have it' and those who cannot.

The point is that, even if you can afford 'surfing the net', this does not make you someone who can express himself or herself. The hopeful statement that 'The

Mirror World [a virtual, software copy of a city, according to Gelertner] isn't snoopware. Its goal is merely to convert the theoretically public into the actually public. What was always available in principle merely becomes available in fact' (Gelernter, 1991, p.19) ignored the danger that the 'public' in cyberspace can be easily, and in fact is, an elite of people who, for several reasons, are more suitable than others to adapt themselves to the virtual world.

Indeed, being visible and empowered is something that implies the ability to participate, taking active part in what goes on in the digital community. As Hoogvelt and Freeman noted: 'Communities online grow from communication rather than information retrieval' (Hoogvelt and Freeman, 1996, p.3). Digital cities 'need to stress interactive communication (through e-mail etc.) rather than just passive information consumption' (Aurigi and Graham, 2000, p.494).

Ken Cottrill, quoting from a survey published in 1995 by the US Census Bureau, gave clear evidence for this, noting that

> White households with incomes above $75,000 are three times more likely to own a computer than households with income between $25,000 and $30,000. The survey reveals that among racial groups, Asian and Pacific Islanders are the most likely to own a home computer, followed by white households. Computer ownership is least likely in black households. Not surprisingly, education correlates closely with computer ownership. Less than 5 per cent of households headed by those without a high school education (the 14 to 18 year-old level, roughly up to A level) have a computer. A household headed by someone with a degree is 11 times more likely to own a computer than one headed by someone who lacks a high school education. The study also found that older households were much less likely to own a computer with a modem. Households headed by over-55s account for a third of the sample population, but only about 15 per cent of home computer users (Cottrill, 1995).

Even if this report dealt with computer ownership, and not with the use of computers themselves, it made clear that it was much more likely that a wealthy, educated and relatively young person could publish things on the Internet, rather than someone who could even access the net, but who had a very hard time in learning how to effectively communicate ideas in cyberspace, because of his/her education level, computer literacy, age. This can lead to an even stronger social polarisation in the city. As Doreen Massey feared, speaking about mobility:

> It is not simply a question of unequal distribution, that some people move more than others, and that some have more control than others. It is that the mobility and control of some groups can actively weaken other people. Differential mobility can weaken the leverage of the already weak. The time-space compression of some groups can undermine the power of others (Massey, 1991, p.26).

What could this mean in terms of cyberspace? It suggests that those who are marginalised in the present society could become even more invisible in cyberspace, by being excluded for economic or educational reasons. It would be a sort of natural selection, that would as well naturally make 'hyper-visible' those who are already in a good social position.

Public administrations wanting to create virtual versions of towns and regions would probably have to deal with this problem if they wanted to assure that as many people as possible can participate in cyber-life. They would have, for example, to consider how to teach the 'culturally disadvantaged' to publish their ideas and at the same time make efforts to increase the user-friendliness of the system itself. A question then raises about this issue.

How have the creators of cybercities been shaping public cyberspace in terms of user-friendliness, and how have they been coping with the problem of computer illiteracy?

Dividing people?

Gender differences could also be crucial in hi-tech environments, that have been traditionally male dominated, as stressed by Katherine Hayles, 'the next time you are in a shopping mall, check out the video arcade. Most of the patrons are teens and preteens. How many are male? If your experience is like mine, nearly all' (Hayles, 1993, p.183). As well, a virtual world with a masculine component and character that is much stronger and present than the feminine one, could definitely be unwelcoming for women, or simply not interesting. Roger Burrows stated that: 'With respect to gender relations in particular there is good evidence to suggest that cyberspace is a sexist male dominated (non) place' (Burrows, 1995, p.10). Again, the fact is that cyberspace is not neutral: it always has a character, and the visitors' ability to interact with it, getting visible and expressing themselves can be heavily affected by what type of environment they can find there.

This suggests that virtual spaces are not necessarily safe from the risk of becoming fortified, protected and somehow elitist places, even if they are offered for free, even if they are not directly controlled by a big corporate company. It is the establishment of a certain group as the dominant community of that place that can discourage others to participate, to get visible, as stressed by Brown when he stated that: 'ethnic groups collect in their own electronic communities, libertarians speak only to libertarians' (Brown, 1994). The myth of the virtual community as a place where differences are naturally overcome and everyone gains an unlimited freedom of speech is evidently flawed.

Further social fragmentation could then be the result of the development of a virtual town, not simply because of the commodification of its virtual spaces, but as a consequence of the self-generation of a series of electronic ghettos:

> Telematically linked communities could fragment our larger society, enabling each of us to pursue isolation from everything different, or unfamiliar, or threatening, and removing the occasions for contact across lines of class, race and culture (Calhoun, 1986, quoted in Graham and Marvin, 1996, p.232).

This does not mean, though, that any attempt to establish public spaces on the Internet has to be considered an agent of fragmentation. How cyber-participation is shaped will determine its effectiveness. As suggested by Bianchini:

It is clear that, by providing opportunities for citizens of different genders, ages, classes, races and lifestyles to meet in a variety of informal ways, social cohesion, mutual respect, and understanding could be enhanced (Bianchini, 1988, p.7).

Policies about gathering places in cyberspace can be crucial in determining how conferences and groups develop themselves, and this can be strongly influenced in the design phase, too. Creating extremely specialised areas in a digital city can possibly lead to fragmentation. But the regulatory aspect can be extremely important in this matter: the policies and their application should aim at managing virtual communities that 'must provide not only a sense of belonging and wholeness for their members, but incorporate and tolerate diversity' (Murray, 1993 – quoted in McBeath and Webb, 1995, p.8). Many choices about language and jargon used in civic cyber-debates, as well as the way information is presented could dramatically affect participation and a particular care should be taken when dealing with these matters.

A further question to be asked, consequently, has been: how have the policy makers been coping with diversity in cybercities? Which technological means have they been using, and how?

The superimposition of cyber and physical

The possible risk of a heavy polarisation of urban cyberspace has been envisaged in the previous sections. This would be characterised by a relatively large elite of 'haves' and 'cans' who can afford to pay for it, to develop the necessary skills and to 'survive' in its groups, and on the other side the 'have nots' who are forced to stay outside of cyberspace or at most to use it in a very passive way.

This can lead us to some considerations about the crucial relationship between physical space and virtual space or, more precisely, between the built city and the electronic one. The access related problems prove in fact that it is at least doubtful that the digital city could become a mirror of the real one, especially in social terms, if careful policies are not applied.

Despite this, the sheer amount of hype surrounding cyberspace has produced claims that tended to downplay social exclusion and to present a quasi perfect parallel between real and virtual world, caused by the integration of the two:

> The division between technology and nature is dissolving as the analytic categories we draw upon to give structure to our world – the biological, the technological, the natural, the artificial and the human – begin to blur (Stone, 1991; quoted in Burrows, 1995, p.4).

The fact that computers are increasingly able to deal with several aspects of perception, such as pictures, sounds, moving images and even tactile sensations, makes cyberspace be seen by some as a more and more accurate model of the reality: 'The more information a computer can process, the more of the world it can turn into information' (*Visions of Heaven and Hell*, Channel 4 Television, 1994). And the virtual copy of the real world, moreover, 'looks safe indeed compared to urban decay' (*Once upon a Time in Cyberville*, Channel 4 Television, 1994). This search for safety and 'perfection' has been the cause in the US of the production of spaces for entertainment that, rather than reproducing some fantasy environments as medieval castles or pirate islands, have been trying to simulate everyday reality as it should be, according to those who shape them. As argued by Mike Davis: 'Today, however, the city itself – or rather its idealization – has become the subject of simulation' (Davis, 1992, p.17). Criticising the creation of City Walk, a tourist attraction set up at Universal Studios in Los Angeles that presented a crime-less and poverty-less version of some real L.A. areas, Davis claimed that

> Designed by master illusionist Jon Jerde, City Walk is an 'idealized reality', the best features of Olvera Street, Hollywood and the West Side sinthesized in 'easy, byte-sized pieces' for consumption by tourists and residents who 'don't need the excitement of dodging bullets ... in the Third World Country' that Los Angeles has become (Davis, 1992, p.18).

'Idealisation' is probably a keyword when it comes to considering how a place that should 'mirror' or even replace reality is being shaped, and the temptation to force things to look better and safer than they are has of course its serious drawbacks: 'I think in fact we already much live inside a "virtual" environment, and what's really being impoverished by this is the world of real experience, and people interacting with each other' (Dewey, from *Once upon a Time in Cyberville*, Channel 4, 1994). And Davis himself warned about those that he called 'urban simulators' noting that

> Indeed, some critics wonder if it isn't the moral equivalent of the neutron bomb: the city emptied of all lived human experience. With its fake fossil candy wrappers and other deceits, City Walk sneeringly mocks us as it erases any trace of our real joy, pain or labor (Davis, 1992, p.18).

The 'simulator' Davis was talking about was a physically built area, with well defined material properties, but it was nevertheless considered something that distanced itself from reality. Being physical does not imply being 'real', at least in social terms, and a urban renewal project that works like a filter, segregating and discriminating entire layers of the society could end up being

> ...an architecture of deception which, in its happy-face familiarity, constantly distances itself from the most fundamental realities. The architecture of this city is almost purely semiotic, playing the game of grafted signification, theme-park building (Sorkin, 1992, p.4).

Urban cyberspace, or at least some of its manifestations, could fit very well into this definition of 'architecture of deception'. Its supposedly 'public' spaces tend to be completely iconized environments. They are even more 'purely semiotic' than those of the physical places described before, and therefore they are extremely suitable for idealisation and for excluding those parts of urban reality that are somehow inconvenient. Moreover, the ability of computers to keep things tidy dealing with large amounts of data in a very short time, has allowed the hype about virtual cities to go further, envisaging a new way to approach and deal with urban space. Going 'out' into the 'virtual version' of your town, instead of using the real environment, not only has been seen as safer, but could give you a feeling of control and 'supervision'.

> When you switch-on your city Mirror World, the whole city shows up on your screen, in a single dense, live, pulsing, swarming, moving, changing picture. This big picture is the 'top surface' of the Mirror World. You can dive deeper to explore, but you start out from the big picture – with the big picture on your screen. When you're finished doing business in the depths, you return to the surface – to the big picture – on your way out. Whenever you use a Mirror World, the image of the whole is available, and inescapable (Gelernter, 1991, p.30).

But, to echo the concerns for the fragmentation of the real city examined in chapter 2, is the 'whole' really a whole? In his enthusiasm, Gelernter failed to acknowledge and to address all the problems related to the shaping of the access to cyberspace, that could be cause of social exclusion. He spoke about seeing 'everyone and everything, without changing out of your pajamas' (Gelernter, 1991, p.23), but it has already been argued that speaking about 'everyone' is utopian and simplistic, and that particular efforts are needed to try and involve a large part of the population in the advanced use of computer networks. That is a very hard task.

It is evident, then, how dangerous it could be forgetting that a computer-based system, such as a digital city, is not necessarily to be seen as a mirror of the real urban situation. Rather, it is very unlikely to be like that, especially if nothing is done to overcome all the problems about the commodification of the information and those related to the ability to access the system and to get visible on it. Henderson acknowledged this and claimed that 'a danger is that the abstraction can become the reality, and real world decisions can be made without due regard to the real world' (Henderson, 1990; quoted in Hayles, 1993, p.179).

This is really a key issue. A cybercity is likely to be an idealised place. But the 'dark side' of idealisation, as we have seen, is 'selection'. Selection of the 'positive' aspects and mechanisms of the city, and sanitation from the inconveniences, exclusion of the 'bad' or apparently useless bits of the urban and social structure. This can, in extreme cases, lead to a quite restricted lobby of educated and wealthy computer-literates to be superficially considered as the 'citizenry', with disastrous consequences for the decision-making processes. It can also lead to the ultimate concealment of what is better not to see, both of the built environment and of the social situation.

What, then, would the techno-citizens who Gelertner was talking about, see on their monitors? What would they be able to grasp about reality from the 'image of the whole'? And above all what would they miss completely, who would they not be able to meet in the 'Mirror World' and hear from them. Which communities would they ignore the existence of?

Virtual, yet as real as possible

Andrew Shapiro focused on these problems speaking about two different possible designs of a digital city. The first, that he called Cyberbia, is a possibly commodified space – definitely a very controlled one – in which reality is highly idealised, and no disturbances can happen as '...you can shape your route so that you interact only with people of your choosing and with information tailored to your desires' (Shapiro, 1995, p.10). He therefore proposed an alternative type of digital city, called Cyberkeley, conceived to guarantee freedom of speech and the presence of public spaces that he defined as 'virtual sidewalks':

> Consequently, it should be clear that Cyberbia – like suburbia – simply allows inhabitants to ignore the problems that surround them off-line. In Cyberkeley, by contrast, people may be inconvenienced by views they don't want to hear. But at least there are places where bothersome, in-your-face expression flourishes and is heard. These public forums are essential to an informed citizenry and to pluralistic, deliberative democracy itself (Shapiro, 1995, p.10).

And these public forums, he argued, could improve communication among people: 'As on a real public sidewalk, a virtual pedestrian can try to ignore what's there and pass right by. Most probably will. But some will be enticed to listen and even to argue' (Shapiro, 1995, p.12).

These claims focused on one single aspect of the whole relationship between real and virtual city, although important, which is freedom of expression. This was probably due to a particular sensitivity of the Americans towards the topic of censorship, further increased by the debate around the communication decency bill proposed by senator Exon and approved in the USA in the 1990s. Shapiro's article dealt only marginally with some of those crucial problems related to the access to the system, that were examined in the previous paragraph. The existence of public forums and virtual sidewalks is certainly important, but it is important as well to determine how many people would be able to be active users of them.

Nevertheless, the idea that 'people may be inconvenienced by views they don't want to hear' is very interesting, to the extent that it stresses the need not to over-idealise the digital city and, rather, to look for a way to link cyberspace to all aspects of urban life. In other words, even if it can sound like a paradox, there was in the 1990s – and there still is – a strong need for 'reality' in virtual space.

The necessity to link together real society and cyberspace was also suggested by Julian Stallabrass when he examined the phenomenon of Bulletin Board Systems. Real communities were seen as the only pre-condition that could justify and underpin the future development of the virtual ones:

> ...the virtual community demands a real one prior to it in order to function successfully. That communities can for the moment flourish is due to the still relatively closed and tecnophile nature of the net, which is just what gives it its coherence, and means that it is hardly the identity-free utopia that some claim (Stallabrass, 1995, p.14).

Claims such as these raise some serious doubts on the techno-enthusiastic visions of a future characterised by a beneficial shift of the physical world towards its virtualisation. Rather, a particular emphasis has been put on the need to shift the extremely virtualised environments towards all those aspects of reality and of personal experience that they could tend to leave behind. If cities need technology, technology itself has been seen to make much less sense without a strong link with the real urban environment.

Re-materialising cyberspace

The relationship between real and virtual is not uni-directional, and if it is true that virtual spaces might be useful to give visibility to physical ones, it is true as well that the physical world cannot be seen simply as an object to be displayed on our monitors. Peter Cochrane argued, for instance, that: '...we will have a transition from the information society to the experience society. It will be about "being there" ... Then we will be able to enter the real world from a distance' (Cochrane, *Visions of Heaven and Hell*, Channel Four Television, 1995). But William Mitchell warned that, anyway, it would be only possible to have: 'Virtual bodies that can sense and act at a distance, but that also remain partially anchored in their immediate surroundings' (W. Mitchell, 1995, p.43).

To what extent could and should virtuality replace the physical experience of the city? In an article titled *Being There*, David Brittan took as an example Chris Turner, from Olivetti Research Laboratories in Cambridge:

> Do you need to see a video image of someone just to be asked out for a beer? 'Well, you don't – Turner admits – but don't you think is rather criminal that you can't?'. In his view, the advent of two-way video on computer workstations is a matter of manifest destiny (Brittan, 1992, pp.43-44).

The Olivetti engineer was not really right when he envisaged the increased use of special, sophisticated two-way video on workstations. Despite this is widely available through the use of webcams and cheap of free software like Microsoft Netmeeting, there is little or no evidence that it is bound to be used heavily within contiguous office environments. But above all his enthusiasm for the replacement of some basic human contacts with a video link made him define 'criminal' the lack of multimedia interface between two adjacent offices. This attitude was probably due to that same phenomenon that McBeath and Webb defined as 'fascination and absorption' saying also that:

> When we are involved in a group conversation in cyberspace we are not in intimate relation with the other persons but with the excitement of anticipation of what is going to come on screen next, or what one is going to say in response, but ultimately always with the magic of the immediate presence of the image into which we are incorporated (McBeath and Webb, 1995, p.12).

Scepticism about the ability of virtual social life to replace more traditional forms of interaction has been shared by many researchers, and the need to return to physical places, or better not to try and escape from them, has been expressed from several points of view. Allucquere Rosanne Stone, for example, noted that 'Cyberspace developers foresee a time when they will be able to forget about the body. But it is important to remember that virtual community originates in, and must return to, the physical' (Stone, 1991, p.113).

In terms of urban life, the ability to go out and meet other people can then be considered an experience that should not be replaced and cancelled by any kind of electronic relationship:

> We argue that public interaction on streets and in public spaces offers much more than can ever be telemediated. Real face-to-face interaction, the chance encounter, the full exposure to the flux and clamor of urban life – in short, the richness of the human experience of place – will inevitably make a virtual community a very poor substitute (Graham and Marvin, 1996, p.231).

Overcoming the limitations of the physical world by living 'electronically', does not seem to be a satisfying answer to the problems of fragmentation and mobility of contemporary cities. Rather, it seems to introduce new, heavier limitations to social life:

> Going out, going to work, going to school or to church, going away to college, and going home are economically significant, socially and legally defining, symbolically freighted acts. To change or eliminate them, as electrocottages and cybercondos promise to do, is to alter the basic fabric of our lives (Mitchell, 1995, p.103).

Or it can be seen as an overestimated form of entertainment, which has been given too much importance in social terms. Here is how McBeath and Webb criticised the hype about the so-called 'Multi-User Dungeons' (MUDs), on-line fantasy games often described as an interesting example of virtual world, constituting: '...living laboratories for studying the first-level impact of virtual communities' (Rheingold, 1994, p.146):

> In the case of Multi-User-Dungeons and Furry-MUCKs, which are part discussion groups and part fantasy games, users are just engaging either with on-line Dungeons and Dragons, or developing in conjunction with others an adult version of Brer Rabbit stories. These are just glorified children's games of make believe. Participation in MUCKs and MUDs may require use of the imagination but it does not secure the affective dimensions of community (McBeath and Webb, 1995, p.8).

It is clear enough, then, that any technocratic view of the future of cities, tending to assign to cyberspace the role of the new, exciting place for human interaction and debate, due to replace the worn-out physical town, is just simplistic. Even if it was possible to overcome the inner pitfalls of a cybercity that were examined before, such as commodification of information, and polarisation due to uneven access and participation, digital cities ought not to be conceived as being capable to create brand new communities, replacing the city itself. The previous claims suggested that, rather, digital cities should be 'physical space-oriented', encouraging people to go out, and helping the 'real' city and the 'real' communities.

A circular relation

The contributions reviewed so far suggest that cyberspace does not seem to have the potential to create communities and consequently an alternative city that is able to compete, in terms of the richness of its experience, with the traditional, yet problematic, urban public space. It has been argued that physical reality is still the basis on which virtuality can develop, especially from the point of view of the individual human being. It has been argued, above all, that the physical city is important to validate cyberspace itself and to give it a reason to exist, a social purpose other than that of being a place for entertainment or a private club for technophiles.

We noticed how Andrew Shapiro recommended that many aspects of reality, even the disturbing ones, could find a place in the virtual city, constituting its 'public spaces'. This is an important point indeed, but this concept should probably be extended to the possibility for a digital city to involve and promote the physical one. This is why Stephen Graham argued that

> The challenge to planners and local policy makers is to try and construct the meaningful, accessible and local virtual cities which support the positive publicly-supported urban vision of trying to re-connect the often fragmented elements of cities together (Graham, 1995b).

And, highlighting the fact that the two dimensions of physicality and virtuality should never be considered independent from each other, William Mitchell saw the problem from the point of view of the architects who are used to deal mainly with physical space, observing that

> Increasingly the architectures of physical space and cyberspace – of the specifically situated body and of its fluid electronic extensions – are superimposed, intertwined, and hybridized in complex ways. The classical unities of architectural space and experience have shattered – as the dramatic unities long ago fragmented on the stage – and architects now need to design for this new condition (Mitchell, 1995, p.44).

These two points of view are in a way complementary and can help us in understanding the circular relation that virtual and physical public space share. If it is true in fact that the vision of cybercities replacing real ones and creating new electronic communities is to be considered simplistic, it is true as well that the ability of cyberspace to give visibility to people and places should not be underestimated, and that it can be a tool to 're-connect the elements of the city' as Graham (1995) suggested. At the same time, design issues about cybercities could involve not just how they are shaped electronically, but – as argued by Mitchell –

how architects, urban designers and planners might decide to manage the overlapping, the 'superimposition' of virtual and physical elements. It is a circular relation because the city goes into cyberspace and cyberspace goes into the city, influencing each other.

Rather than creating communities, a digital city could be used to link the existing local communities together, to give them the visibility that has been lost because of the problems of communication in the city: 'When networks and information servers are organized to deal with information and issues of local concern ... they act to maintain more traditional, site-specific communities' (Mitchell, 1995, p.116). Cyberspace could be used not as a replacement of the town, but as a reference to real people and real places, even those that might have been forgotten or that might be considered either unsafe or not interesting. That would mean recovering through communication an awareness of the richness of the places that exist already, but that have become hidden fragments. These bits of the city could then be referred by public cyberspace, being set up as

> ...visible, accessible and at least occasionally unavoidable. (...) Through regulation of financial incentives, Congress might be able to get users of commercial and academic on-line services to pass through a public gateway before descending into their private virtual worlds (Shapiro, 1995, p.12).

This public gateway that forces users of electronic services to see things they would otherwise avoid, simply because they do not belong to their specific activities, could possibly suggest an alternative type of digital city that focuses on public space, rather than on providing ad-hoc services, as many of the 'cities as menus' discussed previously actually do. And for the cybercity to deepen the relation with the real one, not only people might be promoted, but physical places also, to produce a geographically referred virtual town, the city going into cyberspace and cyberspace promoting the city.

Cyberspace in the streets

Other proposals have added to the debate, stressing the idea of cyberspace merging with the built environment, enriching it with a new communication technology and using the appeal Information Technology has to attract people to the streets again, whilst trying to overcome the access problems for those who cannot or do not want to use telematics at home:

> There is complete dissociation of the two [physical and virtual space] if the electronic public space is only accessible from personal computers in homes and businesses. Another possibility is to associate access points with civic architecture; put an electronic information kiosk in the lobby of city hall, or in the public library, for example. The Berkeley Community Memory and Santa Monica PEN systems have demonstrated the possibility of a more radical strategy by placing rugged workstations in places like laundromats, and at congregation points for the homeless; these workstations thus begin

to play a public role much like the traditional one of fountains in the public spaces of Rome (Mitchell, 1995, p.127).

These reflections can lead to a very different way to look at the relationships between physical and virtual space. Rather than considering these as two separate environments that somehow could even be seen as antagonist or alternative to each other, place remains one, and integrates, combines, or as Mitchell would argue, re-combines, its physical and electronic fragments together in new 'smarter' configurations:

> Building these programmable places is not just a matter of putting wires in the walls and electronic boxes in rooms (though that is a start). As the relevant technologies continue to develop, miniaturised, distributed computational devices will disappear into the woodwork. Keyboards and mouse pads will cease to be the only bit-collection zones; sensors will be everywhere. Displays and effectors will multiply. In the end, buildings will become computer interfaces, and computer interfaces will become buildings (Mitchell, 1995, p.105).

This of course can be applied to cities as well as buildings. The 'recombination' of elements of the built environment and electronic bits can be seen as an ongoing phenomenon:

> We will characterise cities of the twenty-first century as systems of interlinked, interacting, silicon – and software – saturated smart, attentive, and responsive places. We will encounter them at the scales of clothing, rooms, buildings, campuses and neighbourhoods, metropolitan regions, and global infrastructures (Mitchell, 1999, p.68).

Considerations on the recombination of elements of space and cyberspace have even prompted Thomas Horan to envisage guidelines and principles for tackling the design of 'recombinant' places. Horan suggests throughout his book *Digital Places: Building Our City of Bits* that attention should be given to understanding 'digital places' as the 'interface between electronic flows and physical places', and designing and building 'synergistic combination[s] of physical place and cyberspace' (Horan, 2000, p.14 and p.17).

All of this widens the issue of the relation between physical and virtual. Cyberspace going back to the physical can imply not only promoting real spaces and people through telematics, but materialising telematics themselves, and using cyberspace terminals – the doors to the electronic world or, better, to the electronic version of the world – as elements of the real city, making traditional sites value-added places, as Magnaghi envisaged:

> The real squares, of course, will not disappear, replaced by the virtual ones. Rather, we will have to bring cyberspace to the real squares, as a system of communication and additional knowledge. But we will have to recover completely the 'wisdom of the sites' and to produce a new process of settling down: to bring cyberspace to people and not people to cyberspace (Magnaghi, 1995, p.23) [translation by the author].

This relates to this analysis of the digital cities on the web. It is, in fact, another aspect of the design of a digital city, that can moreover have a particular relevance about the shaping of public space. Public space, and above all the public sphere, is in a way becoming a hybrid between physical and virtual, and this happens whether we like it or not, as an effect of the ever-increasing useage of ICT in our daily lives. But when it came to it becoming so in a planned, purposefully designed way, the consequences could be interesting. Access can also be heavily affected by decisions to take cyberspace terminals to public areas of the town, and as it has been said, access is a crucial issue for the whole justification process of a cybercity.

Differences in the shaping of virtual towns had then to be examined not only in relation to commodification and ownership, or to access to the service. How virtual cities were placed in relation to the real, local urban environment had to be studied. Two sets of questions addressed this. First of all, how had the designers of cybercities been trying to establish public debate and to give visibility to local communities, even those representing the deprived bits of the city? And how had they been shaping virtual towns in relation to the fragmentation of both social and physical space, to link the bits together? Had they been referring cyberspace to real space? And how?

Also, had designers and policy makers been trying in some ways to bring cyberspace to the streets, integrating it with real places? And how? How the development processes and the planning processes that are shaping real urban places could relate to those shaping the electronic space?

Issues for a typology

The first part of this chapter raised the issue of the need for a typology of web-based digital cities, in order to move the first, yet significant steps, in understanding the scope and nature of the phenomenon. The reflections produced in the rest of the chapter have helped focusing on a series of main principles that could be used to classify and evaluate urban Internet-based information systems, and that can inform and shape the first phase of the investigation.

Ownership

This point simply needed to be identified for the many digital city initiatives available on the Internet. Knowing who owned and run certain digital cities could – and yet it could not – suggest determinate correlations between digital city performance and the shape and type of ownership.

Informativeness and serviceability

Digital cities in the 1990s could be regarded as a particular breed of urban information system. So, obviously the type and richness of the information provided and – where applicable – services implemented, seemed crucial aspects to observe and analyse. These were basically the 'contents' of the digital city as an information repository and a service platform, and without these it would be very likely that the usefulness of the system itself would approximate zero, ending up being a sort of an empty virtual box. Urban regeneration as well as the empowerement of citizens need effective information provision. Although this would not be a sufficient condition for a 'best practice' case, as indicated below, it would definitely be necessary, and had to be observed.

Social access and participation issues

As extensively noted earlier in this chapter, the ability to access and use the information, services and virtual spaces of the digital city is not something that can be just taken for granted. Several barriers exist for accessing new technologies effectively, depending on levels of economic wealth as well as education and training. Age and gender have also been noted as possible discriminatory factors that could keep people from the fruition of an allegedly public virtual environment, as pointed out in the Interim Report of the High Level Experts Group of the European Union titled *Building the European Information Society for Us All*, where it was recommended that attention should be given towards the inclusion and facilitation of use of 'information society' services by disadvantaged social groups such as 'the elderly, early and "active" retired people, the unemployed and women' (European Commission, 1996, p.35).

It has also been noted how the notion of 'access' to community-related ICT initiatives and digital places needs to be regarded as not just the ability to retrieve information from a web page, but also – and above all – as the empowerment of participating in an active way in what goes on within a digital environment. Access and participation are therefore strictly related, to the extent of being regarded as sides of the same coin, participation being 'active' or 'pro-active' access. This is obviously possible only whenever the technology enables people to communicate, rather than just 'read' on a screen. So a typology of digital cities had to consider the presence of facilities enabling rich bi-directional flows and communication as a crucial factors that could differentiate some initiatives from others that tended to stick to the commercial, elementarily interactive model of 'press button to purchase', or in the case of digital cities 'press button to read the information'.

'Embeddedness': the relation with the host city

These reflections have also highlighted how relevant and important the relationship between the 'digital' city and its physical counterpart could be and, consequently, the ability of the virtual city to address 'real' geographically-based communities and the issues that matter to them, rather than to electronically displaced groups. The choice of this book has been to examine the former type of digital city, the one with a clear and possibly strong relation with its host town, an urban information and communication system that could be defined as 'grounded' (Graham and Aurigi, 1997), or 'embedded'. It has been argued that initiatives of urban ICT that privilege the relationship with the local economy and communities can be a key innovation factor for regenerating cities, rather than making them somehow weaker or obsolete. Organisations like *Telecities* in Europe, or *Communities Online* in the UK were established to allow knowledge exchange and facilitate the development and deployment of 'grounded' projects. Dave Carter, a policy-maker from Manchester and one of the founders of Telecities, highlighted the importance of nurturing local telematics, as 'a critique of what might be termed the "Utopian School" of future "cyber-lifestyles" which sees cities becoming depopulated, 'instant electronic democracy' replacing the need for governmental structures and services and a dominant "ruralist" lifestyle emerging' (Carter, 1997, p.139).

Other commentators stressed 'the advantage of community networks in comparison to other forms of virtual community, which have only intermittent or an absence of shared geographic space' (Wakeford, 1996), where 'community networks' clearly identifies place-based communities able to reap the benefits of telematics and the Internet.

So, an important aspect to survey about digital cities was their 'embeddedness' (or 'groundedness'), and therefore how locally-relevant their information, services, and participation facilities were.

Visions that shape the digital city

In this chapter a wide range of issues that could characterise the shaping of a digital city initiative have been analysed. This review has demonstrated that the black box of the 'cybercity' needed to be opened up and watched closely and carefully. The fact of embarking in the creation and deployment of a cybercity did not imply beneficial consequences or impacts per se, and cyberspace effects and relations with the urban environment could vary dramatically depending on a series of design issues and policy choices. Approaching the study of the digital cities phenomenon should then have implied first of all an observation carried out with a particularly open mind. Different systems could be shaped in radically different ways, so that virtual cities could work for even opposite purposes, and possibly produce opposite results both on the social and urban scenario.

This again suggests how important are the observations made in Chapter 2 on the need to consider the ways technologies are socially constructed. Different approaches towards the same problem, and different ways of seeing the problem or the task itself would inevitably lead to different technological trajectories and, with them, urban futures.

It was crucial here, once the more 'visible' or observable parts of the initiatives had been surveyed, to go beyond that and get knowledge of who the actors were who were shaping these experiments, what visions, needs, and expectations inspired them, how they saw the the city and innovation in it. An approach was needed, linking the more digital city-specific reflections of this chapter with the more general consideration of the social construction of technologies from Chapter 2, and looking at how socio-political and economic factors, as well as interpretations of the city, the role of government, and the role of businesses and citizens, affected the adoption and shaping of new technologies. This required in-depth case studies and more qualitative work, which were the other methodological tool characterising this investigation.

PART II
THE EARLY STEPS
OF THE DIGITAL CITY

Chapter 4

The Web Cities Phenomenon in Europe

Introduction

This chapter deals with analysing the results of the survey work that was carried out to answer a series of 'descriptive' questions about the development of digital cities in the EU. These questions and what was looked at to address them are listed in the following table:

Table 4.1 Survey-related issues

QUESTION	WHAT WAS LOOKED AT
What is the scope and scale of the digital city phenomenon within the EU?	Number of sites per city. Number of sites and density in different countries.
Co-existence of public and private interests. Who owns and/or manages the sites?	Ownership of sites.
How are policy makers relating the d.c. to local economy? What facilities are they providing for local firms?	Provision of web spaces to firms and presence of detailed tourist information related to hotels and other enterprises.
What level of interaction are the digital cities offering to allow active participation?	Overall presence of facilities for interaction and participation.
How are they coping with the tension between earning revenue and social access?	Presence of direct or indirect advertising. Commercial offer of web spaces.
What are they providing in terms of active free access to information i.e. web publishing and participation to debates?	Availability of public forums and discussion areas. Provision of free web spaces for non-profit organisations and/or individuals.
How are they trying to establish/encourage public debate within the city? How are they trying to give visibility to local people and communities, including the deprived ones?	Availability of public forums and discussion areas geared to local issues. Free web spaces for local non-profit organisations and/or individuals.
Are digital cities somehow restricting access? Are they reserving areas to local people?	Areas of the sites with access restricted to local or specially selected people.
What language(s) is being used? Are local people the main target?	Use of local and/or foreign languages (usually English) on the site.
Are digital cities actually providing local information and services, and to what extent?	Provision and overall depth of local information and services. Provision of free of charge web spaces for at least local non-profit organisation.
Drawing from the previous issues, how can we develop a typology of digital cities?	

A typology of city-related web sites in the EU: a key to read the phenomenon

The answers to the questions presented above could also be used to produce scores that would define whether each initiative was 'informative', 'participative' and 'grounded'. The combination of the scores on the three attributes would then generate a typology of city-related web sites in Europe.

It is important to describe the way the typology of initiatives was constructed stemming from the variables/answers analysed. The information collected through the survey's record had to feed the three general criteria adopted to construct the typology. Therefore, nearly each of the variables produced results which, apart from being used to describe specific phenomena by histograms and pie charts, as it will be shown later on in this chapter, had to generate a score determining how 'informative', 'participative' or 'grounded' a certain initiative was.

Some of the variables actually generated more than one single score, as they were considered to affect more than one aspect of the typology. For example, the provision of public forums was an indicator of a participative vocation. At the same time, the local orientation of the forums' topics could indicate the efforts of the initiative to create a strong relationship with the real city and its communities. This meant that the 'public forums' variable generated two different scores affecting two different sides of the typology. The scores were assigned to each option in order to make a maximum result of 100 possible for each of the three typological aspects. Within each of the aspects, the relative difference of 'weight' of each variable's score depended on the importance that that specific variable – and the options within it – was judged to have for the typological aspect it was affecting. To make a clear example – and the most extreme within the whole analysis – although the provision of tourist information and organisational pages clearly affected the 'informativeness' of a certain web site, the presence of reliable public information and services was much more important to determine how 'informative' that initiative was. The lack of a good result within this last variable, even in the presence of the previous two, would have meant that the initiative would be scored as only moderately informative. Therefore, it seemed appropriate to introduce differential 'weights' within the scoring system, and that, in this specific case, the variable 'public services and local information' would produce a much higher score than even the other two added together.

The single scores, once added, allowed the quantification of how informative, participative, and grounded each of the web sites tended to be. It was then decided to recognise a threshold that would allow differentiation between strong and weak trends contributing towards the three attributes generating the typology. In other words, it was decided that, for instance, an informative initiative had to score 40 or over, the same applying for the other two 'attributes'.

The intersection of the results for the three attributes would then determine the 'type' of digital site each of the initiatives was going to fall into. A frequency distribution of the types was then used to produce a typological pie chart showing the general trend in European large cities.

It has to be stressed at this point that the reliability of this scoring system had to be tested somehow, not to produce inaccurate and misleading results. Iteration here was the key to increase accuracy and reliability. Once the typology had been 'calculated', as explained above, a sample check over some of the 'edge' sites (those sites producing scores that were very close to the next category within the typology) was carried out. Around 20 sites were visited again, and the results coming from the spreadsheet were compared with the overall, holistic impression coming from the new visit. This process turned out to be extremely useful to understand that some of the initiatives that had fallen into the 'brochures' category could actually be more appropriately defined as 'civic databases' (the adjacent category). This obviously led to a review of the whole scoring system, and the application of several adjustments to make it more accurate and reliable. Table 4.2 shows how the typology was defined:

Table 4.2 Ideal types of city-related web sites

	Informative	Participative	Grounded
● = Yes ○ = No			
DIGITAL BROCHURE	○	○	
INFORMATION SERVICE	●	○	
Tourist/Investors Kiosk			○
Civic Database			●
ELECTRONIC PLACE	○	●	
Cyber Mall			○
Cyber Square			●
HOLISTIC, URBAN ANALOGY	●	●	
Global Cybercity			○
Embedded Digital City			●

The definitions for the various 'ideal' types of city-related web site were chosen to find effective analogies with existing artefacts or places present in the physical world. So, for instance, a site that scored poorly both for 'informativeness' and 'participativeness' was very likely to end up being an electronic version of an advertising brochure with shallow information of practically no use to the citizens. At the other end of the scale, the 'holistic' types, and especially the 'Embedded Digital City' could represent an attempt, or a trend, towards a fuller use of the capabilities of Information and Communication Technologies for the social and economic regeneration of their host cities.

As stated before and shown in Table 4.1, the typological distribution of the initiatives is the ultimate consequence of the survey work, being generated by the scores derived from the observation of several characteristics of the web sites. Therefore, it could seem logical to show it at the end of this chapter, after all of the single questions had been dealt with. However, the choice here is to show the typology first, within the answer to the first question about scope and scale of the digital city phenomenon. This is because some of the research issues on the topic can be addressed in a much more meaningful way by comparing the general results relative to all of the sites with those relative to the 'Embedded Digital Cities' only. The 'Embedded Digital Cities' were the closest attempt towards the establishment of a civic electronic environment that tended to be useful first of all to the local people and inclusive at the same time. They were the ultimate object of the study described in this book, and it is very important to examine how many they really were, where they were and how and by whom they were being shaped. Although the case studies explained in the next two chapters address more deeply some of the questions relative to the social shaping of two examples of 'Embedded Digital City', the results of the survey are nevertheless useful to:

- Understand how the World Wide Web technology was being used in Europe to create sites that claimed to be useful to the citizens of large towns.
- Understand where the 'proper' virtual cities were located; reflect on possible causal relationships that affected the geography of digital cities, and explore what features these initiatives were actually based upon.

How to look for digital cities?

Surveying web-based civic sites in the EU presented – and would still do right now – a series of challenges, and required a series of choices.

A first choice was excluding those 'web cities' that – although residing on servers physically based in the EU – presented themselves as Internet sites without any special relationship with a real place. Several examples of this type of site can be made, such as 'City Island' or 'Fortune City', which were both placed in England, but completely disembedded from any particular location, sort of 'stand-alone' Internet pages that exploited – sometimes very successfully – the spatial metaphor of the city. This exclusion stemmed from the objectives of this study, as it aimed to analyse how specific digital cities were being shaped to establish a relationship with the urban public sphere of specific towns and cities. The presence of a real urban environment directly related to the web city examined, was a crucial feature here. For this reason, the search for web cities to analyse had been conducted using indexes and directories of real cities on the World Wide Web, and so excluding 'stand-alone' digital cities.

Once it was defined that web sites somehow linked to real cities had to be surveyed, it came to searching for the relevant digital sites. No truly comprehensive index or directory of digital cities existed. This is a quite normal occurrence for the Internet, as pages are born and die continuously. Moreover, unless the manager of a certain information site registers it in some directories or search engines, it is much harder to find it without having previous knowledge about it. This clearly indicated that conducting a really comprehensive, all-embracing survey of web sites of some type would have involved the use of huge human resources, in order to obtain a 'picture' of the situation. Resource and time constraints played a major role in determining the way the surveyed sites were searched for. For the reasons identified above, a selection of the several indexes and search engines of the Internet had to be done. It was decided to rely upon

'City.Net' as the primary source of links to web cities. City.Net was a worldwide directory of web presence of cities. It listed cities by geographical area and/or nation, dividing the sites relevant to each city by categories that appeared particularly helpful for the selection of those to be included in the sample. As the survey dealt with web sites that presented themselves as information sites mirroring the real city and supposedly useful to the citizens themselves, all the pages listed under the 'Travel and Tourism' category were not considered. Exceptions were made for some sporadic cases in which the sites had some very explicit words in their titles, such as 'Virtual City' or 'Virtual Community' etc. This was necessary because there could have been a mistake in the categorisation by City.Net, and some of those sites could actually have been of interest to the research.

Additional sources of information were also considered, and the survey did not rely solely on City.Net. 'Yahoo' was also used in conjunction with City.Net. This was to deal with the fact that some administrators of civic sites could have chosen not to register with the latter. It was a good choice, and in several cases gave significant results, and helped to identify the sites that otherwise would have been overlooked. In some cases, to further increase the accuracy of the search, national directories of civic web sites were used, when known. This happened for the case of the Dutch Digital Cities page (which used to be at http://www.dhp.nl/) and for the CCTA Government Information Service in the UK.

Mainly due to constraints of time, it was decided to consider only EU cities having about or more than 200,000 inhabitants. The 'about' here is necessary, as consistent data about the population of EU cities could not be found, either in atlases and other publications. This search of EU publications and atlases resulted in a list of 167 locations in 14 European countries (Luxembourg does not have any big cities).

Another limitation to the sample came for cases of cities having more than three digital counterparts. It was decided in fact to limit the analysis to a maximum of three web sites per city. In the relatively few cases where more than three sites were present, it was made sure to give a priority in the selection to those that looked more complex, or that were presented as 'official' sites of the city. It seemed that these limitations would lead, however, to the collection of data on at least 200 web sites. The actual final result confirmed this forecast, as 213 city related pages were included in the survey.

Scope and scale of the web city phenomenon

The first question to address, and the main one as well as far as the survey was concerned, was about the scale of the phenomenon. How many web sites could be found between the end of 1996 and the beginning of 1997 in the larger cities of the EU that were claiming somehow to be there as public spaces and services for the citizens?

The first results of the survey were extremely meaningful, as they indicated that the phenomenon of 'urbanising' the World Wide Web was already quite vast.

Within 167 cities considered for the survey, 213 working web sites were found, which meant an average density of 1.27 city related Internet sites per city.

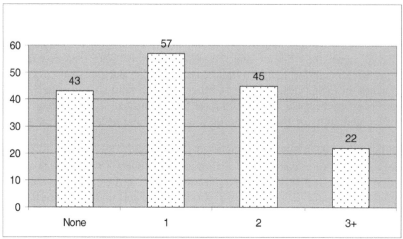

Figure 4.1 Civic web sites per surveyed city

If we start looking more in detail at the numbers in Figure 4.1, we can realise that these web sites were well distributed, leaving only 43 cities (which is slightly less than 26% of the total) without a fully working initiative. Moreover, more than half of the cities that hosted some kind of civic web initiative, actually had more than one site offering Internet based information and services about the place itself. If we also consider that all the tourist oriented only sites had been excluded from this particular study from the beginning, we can understand that the phenomenon was numerous and important, and that the Internet was already widely used to set up information sites about cities.

Could it then be argued that the digital city movement was somehow a mature phenomenon and that informative and inclusive civic digital sites were being established as tools for social and economic regeneration of Western European cities? Not quite. If we just go beyond the mere basic figures and associate them with the results of the typological classification of the initiatives.

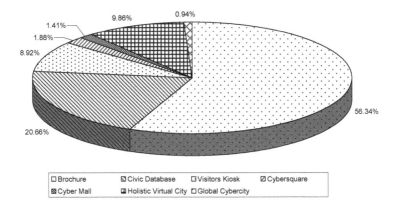

Figure 4.2 Typological distribution of web cities

Figure 4.2 shows that less than 10% of the total of the sites could be considered as 'Embedded Digital Cities'. The chart reveals also that among the surveyed sites, there was a general lack of 'participatory' features. In fact, if we sum the values of the last four categories, those that were providing facilities to encourage participation and expression, we can observe that less than 15% of all the city related web sites were showing in their design some concern towards public participation and discourse. On the other hand, a vast majority of over 56% of the surveyed initiatives was definable under the label of 'Digital Brochure', while another 20.66% was formed by 'Civic Databases'. This overwhelming prevalence of sites that simply provided mono-directional communication, together with the fact that most of these sites had very little, and often not very useful, information, showed quite clearly that the Internet was used mainly as a cheap publishing medium, with overall advertising or propaganda purposes. This type of attitude that made potentially innovative initiatives – or at least using an innovative medium – stick with old models, could be referred to the findings of Guthrie and Dutton, and to their definition of 'broadcasting paradigm' guiding similar initiatives in the US. As they argued: 'Implicit in this paradigm is the idea that the owners of the communication channel control information content and broadcast what is beneficial to them' (Guthrie and Dutton, 1992, p.584).

In summary it can be argued that the phenomenon of the urbanisation of cyberspace was wide but shallow, and that few initiatives were designed to encourage and support public discourse and a wider access of citizens and firms to the benefits of the Information Society.

Geographies of web cities

Another relevant set of results that helps defining better the scope of the web city phenomenon in the EU is the distribution of sites per country. Figure 4.3 shows the general results by country in terms of absolute values – that is how many sites were surveyed – as well as density, as some smaller or less dense countries naturally have fewer cities bigger than 200,000 people.

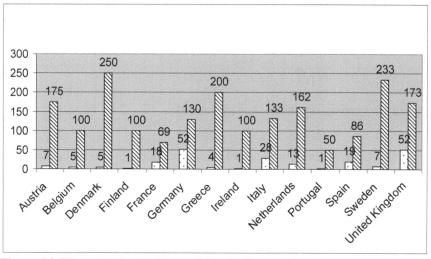

Figure 4.3 City-related web sites and density by country

The figures show that although the majority of the Northern European countries hosted more city related sites than their Southern counterparts – which was a somehow expectable phenomenon – the divide between North and South of Europe was not very sharp. In fact, among the countries that had a higher density of sites such as Sweden (2.33 per city) or Denmark (2.5 per city), Greece had a comparable position with 2 sites per city. Italy did not perform badly at all either, with 28 surveyed initiatives for a density of 1.33 sites per city which was actually higher than Germany's one (1.3 spc). A case apart seemed to be France, which had a relatively small number of city related sites and one of the lower densities in Europe with respect to the number of large cities present in the country. Although this could strike as something unexpected, as France's long standing involvement towards technology innovation and telematics in particular was well known, perhaps the phenomenon can be explained by considering the influence that already existing telematics applications such as Minitel could have over new developments in that country. The diffusion and widespread use of Minitel in France, due to facilitating policies not adopted in other European countries, was probably reducing the need for more sophisticated electronic environments on the Internet. Minitel can be an extremely slow communication system, and it was obviously much less graphically appealing and versatile than the World Wide Web.

Despite this, the well established familiarity of the French public with the service, together with the very low costs involved in using it, were probably making it more than enough for day-to-day information broadcast and interpersonal communication.

As we have already seen, whilst the general figures seem to show a relatively well developed phenomenon, being able to consider the outcome of the typology-making work helps to assess the situation more deeply. This is even more important when we consider the distribution by country of the 'Embedded Digital Cities' only. Figure 4.4 shows how the supposed divide in the use of the Internet for urban regeneration purposes between North and South of Europe weakened further. The facts that Italy, as well as The Netherlands, had nearly a quarter of the total digital cities in Europe at the beginning of 1997 (the latter with an outstanding density of 62%), and that the UK, for instance only had 3 virtual cities out of 52 surveyed web sites, tell a different story.

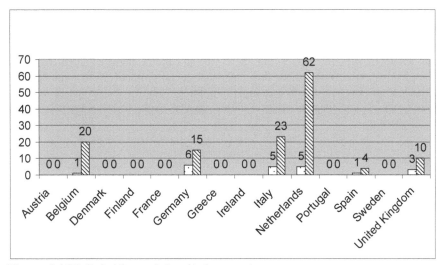

Figure 4.4 Digital cities and density by country

Again, the fact that the Internet was apparently used to provide 'civic' electronic sites does not mean that this was actually happening anywhere. Moreover, these results by country indicated that most possibly, different political and economic conditions were strongly affecting the way the Internet was being used to design and shape city-related sites. The findings relative to the next research question help us to explore some of the connections between political and economic agency and the development of digital city initiatives within Europe.

Ownership of digital cities

Which agencies were promoting the digital city phenomenon? A question about agency was extremely important at this stage, though agency related matters are examined much more deeply in the case studies. However, looking at who was mainly involved in the shaping and running of city related sites could help recognise some trends as well as starting addressing the problem raised in the literature review about the possible 'commodification' of virtual spaces by private companies. Was the presence of private forces going to have negative consequences for the role of cyber-cities towards public discourse, access and aggregation? If this were to be found to be the case, should civic cyberspace be regulated and run exclusively by public sector actors?

Although giving a sharp and definite answer to this type of question seemed impossible, given the relative youth of the phenomenon of civic cyberspace, the findings of this study seemed to suggest that it was not possible to make any generalisation about a clear difference between the ways different actors were shaping their sites. Prevalence between public and private sector in creating city related web sites could not be recognised either. In Figure 4.5, the result relative to the whole of the EU shows that there was a balance between sites that were completely private and others that had been set up with at least some 'public' participation. Even more interestingly, Figure 4.6 shows that when it came to seeing who runs 'proper' digital cities, a very similar situation was being highlighted.

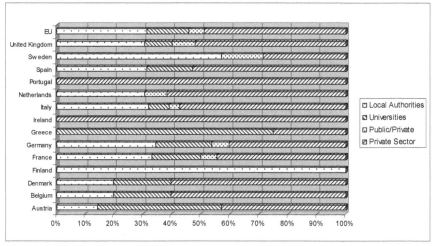

Figure 4.5 Who runs the web cities

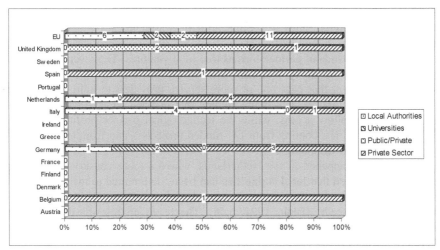

Figure 4.6 Who runs the digital cities

This demonstrated that, at the European level, it could not be argued that the initiative of privately owned companies or organisations would necessarily produce a more shallow and exclusive urban cyberspace, when compared to what public authorities or universities did. Indeed, many local authorities were running sites of the 'brochure' and 'civic database' type. Therefore the dualism between city-marketing interests and the benefits for the actual local residents, highlighted in Chapter 2 for physical cities and policies, was mirrored in cyberspace. Many electronic spaces ended up being used by local authorities and universities very often just as a way of 'selling the city' to potential tourists, investors and students, or at most as a way to broadcast carefully selected and relatively shallow information.

However, going beyond these average European results again helped reveal a series of crucial differences between countries. Still by looking at Figures 4.5 and 4.6 we discover how the same type of actor behaved completely differently in different countries.

The differences in ownership of digital cities in those four countries that had more than just one example of the 'top' type were extremely sharp. Whilst Germany reflected rather precisely the average European results, the other three countries saw a prevalence of a specific 'owner' of digital city initiatives. In Italy local authorities seemed to be extremely active in promoting complex civic sites, to the extent of being nearly the only actor taking care of establishing digital cities (or 'civic networks' as most Italian councils prefer to call them). This was after all not very different from what happens with policies concerning physical urban space, as exclusively public intervention and management of space is still central in a country where councils do not have to share the decision making process with the private sector. Therefore the majority of the holistic digital city initiatives in Italy saw in fact the local councils as the main actors and decision makers on the fortunes of 'public' cyberspace. Not that the private sector was completely absent from the picture, but the nature of many of the so-called partnerships that could be found in Italian digital cities was often limited to a technical collaboration from some IT firms, that did not participate in the policy making process at all. It is hard to describe these kinds of partnerships as true joint ventures. They were based on a precise and strict hierarchy that kept the local council at the very top of the pyramid, as a client that paid one or more service providers to help develop and manage its own initiative. It was the council that decided what to publish on the site and who to allow access, while the technological partner profited from the initiative – sometimes coming from EU funded projects – and from the chance of experimenting new technologies.

The inverse happened in the Netherlands, where a more conservative way of government implied that local authorities had been 'determined to make the public sector retreat from what they see as its exaggerated aspirations to regulate society and care for its needs' (Blair, 1991, p.48). The result of this was that the ratio between public and private digital cities that was observed in Italy was reversed in favour of the latter.

Another aspect that can help to explain the prevalence of privately run digital cities in the Netherlands can be related to the fact that earlier experiments in technology could become paradigmatic 'catalysts' for further developments. In the Netherlands, in fact, a sophisticated digital city initiative was born in 1994, De Digitale Stad (DDS) in Amsterdam. DDS was originally subsidised by the Dutch government, but soon it had to rely on advertising revenue and space rented out to commercial and non-commercial information providers (Brants et al., 1996, p.241). This attitude of the governemnt to 'leave on their own' this kind of initiatives echoes what Bekke has defined as a second form of privatisation in the Netherlands, where

> Initially the government might retain some form of involvement in terms of regulating finances, which would diminish over time. One could think of entrusting swimming pools, libraries and socio-cultural services to companies or private institutions (Bekke, 1991, p.127).

Although a progressive 'detachment' of DDS from the reality of Amsterdam could be observed, for primarily commercial reasons, this experience, and the ways it had developed, strongly influenced the development of digital cities all over Europe, and especially in its own country. It had been standing as a model to replicate or at least to take into consideration, setting a rationale for the development of private, often non-profit but self-sustaining digital cities. This, as well as the likely influence of Minitel in France, mentioned before, indicated that the relation between technologies and society is actually circular. Initiatives can be influenced and shaped by socio-political and economic factors and also existing initiatives can stand as a 'precedent' and affect the ways certain policy related decisions and choices are taken.

The situation in the UK was another example of how context affected the development of digital cities. First of all a consideration has to be made that, despite being the country with more city related web sites in Europe, together with

Germany, the amount of actual digital cities was low. This could possibly be explained by the fact that neither local authorities nor private organisations generally played a role in creating high profile civic sites by themselves. These actors were involved, but the fact was that in the UK the emphasis was mainly put on the power of partnerships to create socially and economically viable initiatives. Councils did not see the 'mission' of creating and maintaining inclusive and communicative urban 'bits' of cyberspace as their own exclusive vocation, as tended to happen in Italy, and had indeed very little money and resources to put in it. Therefore their choice was, in the majority of cases, to use the World Wide Web as an information service or an electronic brochure often managed by their Economic Development sections of planning departments. As we observed for the Italian case, here the condition that local authorities were in control of the agenda for the management of the physical territory was mirrored. And it tended to be a very different situation indeed, compared to that of Southern European countries.

As we have already noted in Chapter 2, the fragmentation of agency in the management of UK cities is very pronounced, leaving local councils unable to play a hegemonic role in the decision-making processes affecting spatial, social and economic regeneration of the city. It can probably be argued that the slow start of UK's digital city initiatives, due to the more troubled establishment of actual partnerships, could be a sign of a still relatively low compatibility and synergy between the interests of the private and the public sector in running and managing cyberspace. At the same time, the situation of the UK looked like a very interesting benchmark for the long term future of digital cities just for the same reasons, as social inclusiveness, economic sustainability and marketing ethos had to co-exist in the same initiatives.

Relationships between city-related web sites and local economies

Observing the relationship between a digital city initiative and the local economy of its host town was certainly a complex task that could not be fulfilled by simply looking at the web site of the initiative. The strategies and choices that, for instance, the economic development department of a city council planned to make in the medium and long term through the digital city were not necessarily shown by the current situation of the site. However, the presence of certain facilities could still be an indicator of the level of attention that who was running a city-related web site was putting towards the matter. And that level of attention seemed low in the surveyed sites.

From the point of view of the broadcasting of information useful to local businesses, only 8% of the sites tried to provide a relatively extensive coverage, whilst 63% of the city-related web sites did not offer any economic and business information at all, as shown in Figure 4.7. 29% of the sites tended to offer what has been defined in this study as 'just descriptions'. That was a type of information that did not need to be updated often, as it aimed mainly at describing how well organised, competitive and popular among enterprises that place was, so as to try to encourage investment from the outside. Again, the 'brochure' paradigm that had driven many of the initiatives was revealed here. As we have already noted, the dualism and tension between the need to encourage residents' interests and the need to market the city – or aspects of it – to external revenue sources was present in digital initiatives as well as it is in physical planning and city management.

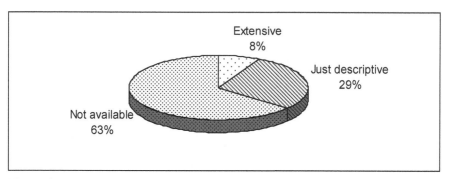

Figure 4.7 Economy and business-related information

Comparing the results shown by Figures 4.8 and 4.9 can also help to recognise this tension, as most initiatives were much more concerned to attract tourists to the city, through offering the relative information, than to provide web spaces. These spaces – either free or charged for – could have represented an integrated and recognisable marketplace and showcase for the host town's firms, as well as a chance for many small and medium enterprises to be encouraged and perhaps trained to start exploiting ICTs in their business. However, 74% of the surveyed sites did not consider hosting firms or organisations. Even as far as the tourist information was concerned, hotels and other partly tourism related businesses such as restaurants, had some kind of presence (often just an entry in a list) only in 34% of cases.

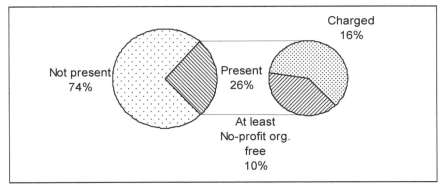

Figure 4.8 Web cities and provision of web sites to organisations

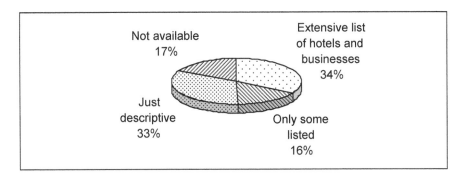

Figure 4.9 Web cities in the EU: tourist information

Therefore, it seems that in general the supposed 'urbanisation' of cyberspace by city related web sites did not try to create particular opportunities for the local economies of the host towns. It often failed to encourage the use of Information and Communication Technologies by local enterprises. From the point of view of data retrieval, little and poor information useful to firms was available, while even less sites had been designed to 'let businesses in' and give them visibility in a sort of a virtual square or marketplace, where they could actively use the Internet. The relationship between these Internet based initiatives and the local economies was basically weak. Among those few sites classified by the survey as proper digital cities, 89% offered web spaces for firms and organisations, still leaving a significant 11% of complex, sophisticated initiatives completely indifferent to relating themselves with third parties and trying to provide some type of virtual space for local enterprises.

Level of interaction offered to encourage participation

This aspect of the sites, which will be detailed in the section about specific participative features, is summarised by Figure 4.10. The chart gives a very clear picture of the situation as it was at the time of the survey. Again, the fact that 73% of the city related web sites were offering at most the chance to write comments to an anonymous 'webmaster', tells a lot about the wide adoption of 'broadcasting' paradigms instead of more bi-directional and participative choices. Writing to the 'webmaster' was used normally to make comments on the web site itself and perhaps suggest improvements, but was obviously useless to create real communication at the urban level. Another 14% of sites published multiple e-mail addresses, for instance e-mail addresses of city councillors and representatives, allowing users to communicate with a limited set of people. This measure also meant that only those users with a valid e-mail account could send mail and express their views. Therefore, the e-mail only facility was limiting the numbers of both recipients and of the potential senders, discriminating against those who could have accessed the web site from a public computer but did not have their own e-mail. Further, e-mail messages are usually private (unless posted in a 'newsgroup'), and this meant that e-mail only facilities could establish a one-to-one exchange of ideas that would be completely missed by other people. This left only 13% of the sites providing some advanced, more sophisticated and inclusive interaction, such as debate areas and newsgroups, real-time chat areas and mailing facilities for those who did not have e-mail accounts of their own.

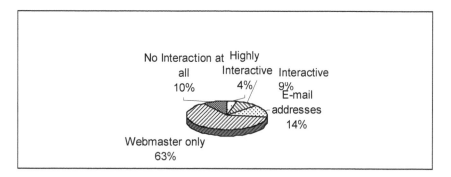

Figure 4.10 Web cities in the EU: level of interaction

Therefore it can hardly be said that web based civic information systems in Europe tended to be communicative, failing often to encourage exchanges between users and more generally bi-directional flows of information. The overall trend was towards mono-directional broadcasting, with the facility to call a single, unknown, operator to leave messages about the service. In substance, it looked like an adoption of a model much closer to cheap television or radio than to 'Rheingoldian' virtual communities, despite the hype often used to present and promote the phenomenon.

Tension between revenue generation and social access

We have just seen how few city related initiatives on the web tended to foster an active relationship with local economic actors. Most failed to make full use of new Internet technologies to encourage or support local economic development. What about the sites' own economies then? Were the web sites trying to cope with their own economic sustainability by generating revenue, and was this going to affect directly the chances of economically disadvantaged people and organisations to have access to these systems? World Wide Web sites could generate revenue in two main ways: hiring advertising spaces – often in the shape of banners somewhere in the pages – and hiring disk and web space to customers who want to publish their own pages and web sites. The survey looked for these features in the sites to determine whether information was directly or indirectly paid for.

The charts in Figures 4.11 and 4.12 show the percentages of sites using advertising or sponsorship banners, or simply hosting firms' names for some reason and possibly getting some revenue out of it. The general results in the first chart demonstrate quite clearly that a huge number of city-related sites still did not bother with trying to make money directly from the web. Some were probably relying on some other type of funding, as it happened for several local authorities' sites getting money from the participation in some European Community project. Others, such as a large part of the 'brochure' sites, were probably far too cheap to run to even need some type of income. Indeed this seems to be one of the reasons for the relative lack of interest in making revenue. It has to be specified that web sites offering advertising space were included in those using banners etc., even if they had not got any actual advert yet. If we look at Figure 4.12, that represents the same type of results, but limited to the complex, proper 'digital cities', we can see that a much higher percentage of sites was looking for some type of economic self-support. Many of these more sophisticated sites were surely facing higher and more frequent expenses, given the level of services offered, and could not afford not to think about getting some money back, mainly through sponsorship.

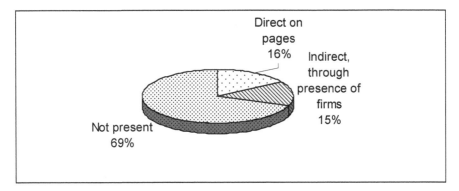

Figure 4.11 Web cities in the EU: advertising

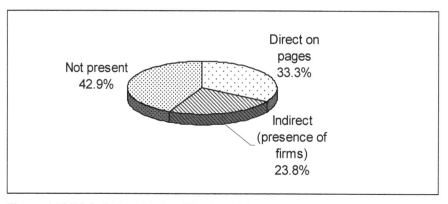

Figure 4.12 Digital cities in the EU: advertising

Confirming the trend highlighted by Figure 4.11, Figure 4.8 shows that the same percentage of 16% of sites that were using direct advertising applies to those selling web spaces to firms and organisations, even if the actual sites could be different in the two cases. Another 10% of sites could well be selling spaces for commercial purposes, but were offering free of charge hosting to non-profit organisations, showing a sensitivity towards the problems of social access, at least of community relevant organisations. The second chart in Figure 4.16 also shows that all of the commercial efforts were being aimed at firms and organisations, but not at all at single individuals. In the very few cases where initiatives were trying to shape themselves more as virtual communities, offering individual web spaces, this facility was not charged for.

In the light of these results it can be argued that the tension between earning revenue and granting social access was not a big issue for the early city-related web sites in Europe. This was mainly due to the poor sophistication and complexity of the majority of the initiatives, which were not aimed at becoming commercially oriented. At the same time, as we will see better in the next section, access was for the same reasons granted but scarcely useful. Other initiatives – by local authorities mainly – were still not looking for direct revenue, offering the digital city as a service paid for by taxpayers' money and European funding. However, very few cases offering more advanced services and facilities did try to cope with the tension, and resolved to do it by subsidising the free access of individuals and non-profit organisations by selling advertising and publishing space to commercial firms.

Free access and participation

The importance of letting people actively use ICT facilities for bi-directional expression has been clearly highlighted in the critical literature review. However, the set of survey results that we have so far examined already suggests that few civic web initiatives were taking this need into account, and that mono-directionality was prevalent with respect to the provision and management of bi-directional communication facilities within digital cities. To have a more detailed account of the extent to which public visibility and debate were allowed and encouraged by the several sites, we can look again at Figure 4.8 and consider the results shown in Figure 4.13. As we have already noted before, the offer of free web spaces for self-representation was relatively poor for organisations and even poorer for single individuals. In fact, only 4.23% of the surveyed sites – therefore a smaller percentage than the 'digital cities' themselves – offered free of charge pages that could be used by people to publish more or less whatever they wanted. Non-profit organisations, often playing an important role within local communities, were just slightly more considered by the civic web sites, as only in 10% of cases they were offered the chance to publish pages for free (see Figure 4.8).

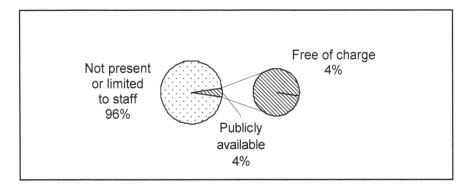

Figure 4.13 Web cities in the EU: individual pages

If the publishing side was neglected by most initiatives, the support of public discourse and debates did not obtain more consideration. As Figure 4.14 shows, only 9% of the sites implemented some type of forum or debate area, where users could leave publicly readable messages or reply to some left by others. This was evidence that very few digital civic initiatives were designed and run to actually provide citizens with an extension of the public realm of their cities in the shape of electronic meeting and debate spaces. But this consideration must be complemented with another piece of results, which is the answer to the survey-related question about support of public discourse at the local level, on local topics.

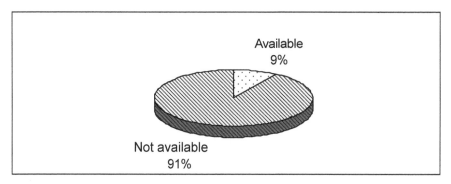

Figure 4.14 Web cities in the EU: public forums

The local 'dimension' of Internet publishing and debating was an important aspect to consider when it came to dealing with digital cities. Their ability to be 'grounded' in the place they were supposed to belong to could be seen as a crucial feature to affect the regeneration of the 'real', physical city. For example, it has been noted that the most famous and paradigmatic digital city of Amsterdam, called De Digitale Stad (DDS), had been gradually tending to offer more non-local discussion platforms, because of the need to attract foreigners and with them 'blue chip' sponsorship. This had resulted in shifting its role from being locally oriented to being more a 'placeless' virtual city, thus weakening its relation with the 'real' city of Amsterdam and its inhabitants (Graham and Aurigi, 1997, p.41). The extent to which web city initiatives tried to revitalise the public sphere of the real city would then be 'measured' by looking at the provision of truly locally oriented debate areas and publishing spaces.

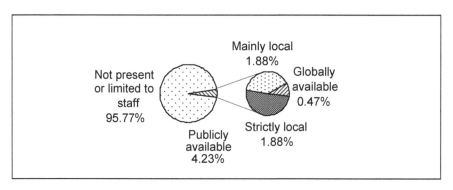

Figure 4.15 Web cities in the EU: individual pages and local targets

This means, as shown in Figures 4.15 and 4.16, that the already low percentages of sites doing something to support public discourse and self-representation decreased further. It was especially the case for debate areas, where one third of the 9% of sites offering them was catering for 'general' non-local topics. Only 6% of the initiatives, then, were offering discussion areas geared at local issues, still often sharing resources with more general debates. On the other hand, the majority of the very few sites providing free web space for individuals ended up offering a free service to locals, rather than trying to attract a more global type of community. It has to be noted that this seemed indeed more a consequence of the pre-existence of more attractive 'global' services on the Internet.

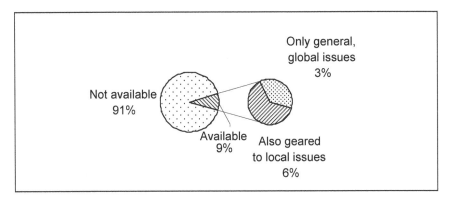

Figure 4.16 Web cities in the EU: public forums and local targets

The presence on the Internet of already well established and organised non-grounded virtual places hosting people from around the world – such as for instance the enormous 'Geocities' site – made a similar offer from smaller sites redundant and less visible. This obviously meant that in offering publishing space to individuals, it was pointless trying to compete with these well organised enterprises at the global level. In a way, the supposed local character of the surveyed initiatives, made them unattractive and relatively meaningless as 'virtual homes' to people from other places, who wanted to be 'seen' and listened to by a global audience. On the other hand, the bonus for local people was evident in addressing local audiences and identifying the virtual space with their physical home. This explains why, in chart 4.15, there is a 1.88% of initiatives being strict on offering web spaces to local individuals, and at the same time another 1.88% (a relevant percentage within the total 4%) not imposing this kind of limit whilst not actively trying to cater for non-locals.

This seems to suggest that the few advanced digital city initiatives that were trying to support public debate and self-representation were ending up doing it with a local, grounded vocation. However, sometimes the choice of general topic oriented debates was not helping to link the electronic communication possibilities with the problems of the physical city. In a way it could be argued that digital cities, or better the few sophisticated ones, were trying to attract mainly local individuals, but failing sometimes to effectively deal with local debates. More generally, however, very few sites were actually addressing the problem in some ways. The majority of them was completely ignoring the possibility of creating electronic extensions of the public sphere of the host physical city, giving people a

chance to be locally visible – at least in the electronic environment – and have a
say in a series of public debates.

Access restrictions

A measure that digital city initiatives could have adopted is to restrict access to
their whole sites or to specific parts of them. It is technically feasible in web sites
to restrict access by password or 'domain' – for instance allowing in only users
who are connecting from computers belonging to a specific group. This could be
done for example to reserve areas to certain categories of users. At one end of the
spectrum, in a way, there could have been access restrictions based on charges for
information, so that a certain digital city could have opted for asking its users to
subscribe to the information service. This would have lead, as noted in chapter 3,
to an increase of the social polarisation between the economically advantaged
subjects and those who were not. But access restrictions could have been used for
other purposes as well, perhaps working in a somewhat opposite direction.
Creating areas reserved to residents only would have shown a concern to run a
service potentially faster for the local people, thence less expensive, and definitely
locally oriented, within that local-global dualism in urban management that we
have already noted about several times. Figure 4.17, however, shows that both
these options were not taken into consideration by nearly all of the sites.

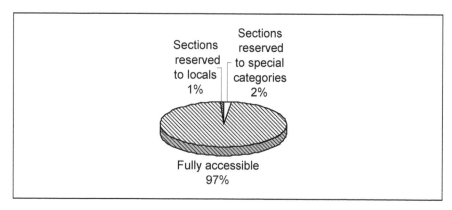

Figure 4.17 Web cities in the EU: access selectivity

It could be said that this is an expectable result, as the choice itself of using the
Internet and the World Wide Web for setting up a civic digital site, rather than
implementing more technically 'local' solutions like a 'BBS' style system, shows a
clear will to 'go global'. Although this consideration is certainly true, we have to
consider nonetheless that the web was more and more becoming a de facto
standard for a multimedia-capable distributed information and communication
system. In fact, several pre-existing community information systems created as
BBSs were gradually 'migrating' to the web because of its better and more flexible

interface. Therefore there were going to be more and more community oriented web sites finding themselves facing the policy choice of 'opening' to the world – and perhaps offering a worse, more congested service to their theoretically targeted local users – or keeping access to areas, or the whole site, restricted. These results show that, at the moment of the survey, only 1% of them had decided to implement some sort of 'locals only' pages, whilst the overwhelming majority had decided to exploit the advantages of global presence potentially offered by the World Wide Web. At the same time, making people and firms pay to view the information was not seen as a fair or feasible option, and at least 98% of the sites remained completely free to consult.

Use of languages

Another factor affecting the possible participation of local people in the 'life' of the digital city was the use of local or foreign language. Again, the tension between local interests and global role of the Internet was central here, and it was important to observe how city related web sites were being shaped from the point of view of language use. Figure 4.18 shows how the surveyed sites dealt with languages, and generally indicates that only 9% of cases were trying to 'speak' exclusively to a global audience, ignoring the needs of local people and especially of those who were not so well educated to be able to speak English. English has to be mentioned as 'the' foreign language used by the majority of the sites, while other languages were much less implemented.

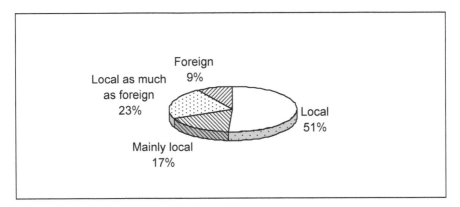

Figure 4.18 Web cities in the EU: languages used

However, Figure 4.18 shows that while 51% of the initiatives were exclusively using their local language, a significant 40% were implementing a foreign one (as already noted, usually English) to address a global audience. However, this phenomenon could also become – perhaps in many cases involuntarily – a means of welcoming foreigners living in the city, thus enhancing the multi-ethnic potential of the sites. But this has to be seen as almost an involuntary phenomenon, as English only was the omnipresent foreign language. Also, if we observe what was happening to those sites where English was the local language, that is UK and Irish initiatives, we find that very few had other languages, and no relevant immigrants' ones were actually implemented, like for instance Bangladeshi or Hindi.

Indeed, the results shown in here could be somehow 'corrected' by excluding British and Irish sites, which had the convenience of naturally speaking the global language, and therefore were improperly – at least in some cases – raising the number of the sites implementing only local idioms and therefore addressing local audiences. This leads to the chart of Figure 4.19, where the 'local only' option lowers significantly from 51% to 36% and very probably reflects more faithfully the actual trend in European web sites.

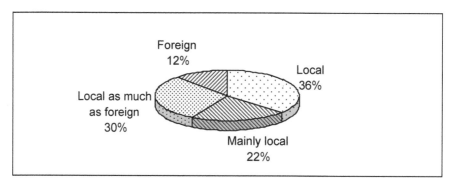

Figure 4.19 Web cities in the EU: languages used, excluding UK and Ireland

However, still only 12% of the initiatives were discriminating against local users or at least the not well educated ones, while the others were mainly (or equally) working in the local idiom, making it possible for citizens to consult them.

In substance, it can be argued that as regards language, local people were the main target, even if the majority of the sites was concerned about attracting foreigners as well. The fact that nearly all of the initiatives using also foreign languages were implementing only English, and in some cases cover only a small part of the pages present in the site, is significant. It tells that the aim was attract foreign investment and tourists from the rich Western countries, rather than setting up information and communication sites that could be more easily used by the immigrant communities that were already part of many multi-ethnic European cities.

Provision of local information and services

We have seen that when it came to constructing a typology of the initiatives surveyed in this study, the amount and quality of information provided were a crucial characteristic to distinguish complex digital cities from more 'brochure-like' systems. We have also seen that another important issue was how 'grounded' this information was, as otherwise it became pointless to speak about relationships between real and electronic places. Any web site distinguishing itself as the digital counterpart of a certain urban place would stop making sense if it did not deal with its host place at all.

Indeed, the presence of more or less sophisticated and complex information was an extremely important set of results to analyse, especially if we consider that many of the surveyed sites ended up being not really 'communication-oriented' and mainly tools for broadcasting data. What these tools were actually being run for, was also revealed by looking at the amount and quality of information and services offered to citizens and users. Figures 4.20 and 4.21 distinguish between two different types of information present in the surveyed sites. Basically, a distinction was made between the presence of information 'broadcast' by the actors who run and control the initiatives – we could call it 'centrally' broadcast information – and the presence of guest information providers such as organisations and firms, participating in the life of the digital city. These two types of information provision were examined differently, as it was assumed that the third party pages belonging to organisations and firms were going to be quite narrowly focused and for this reason exhaustive. Therefore what was checked was their presence and how locally oriented they were, as external information providers could have been for instance big corporate firms advertising globally available services with little relation with the local communities' life. On the other hand, it had been noticed during the pilot phase of the survey that centrally managed local information could vary between being very shallow – almost non-existing – or quite deep, and therefore the quality and quantity of it had to be observed and not taken for granted. As stated, these observations lead to the results in Figures 4.20 and 4.21.

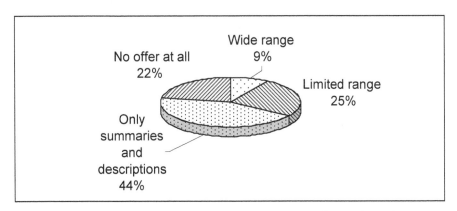

Figure 4.20 Web cities in the EU: offer of local information and services

Figure 4.20 shows the general shallowness of the contents of the web cities in terms of local information provision, as 66% of the surveyed sites ended up not offering information or just providing what in the chart is defined as 'summaries and descriptions'. Information of this kind was made of generic descriptions of aspects of the city, mainly geared at advertising that urban space as particularly suitable for investment or tourism. For example, certain sites could give information about business in the city just describing how many profitable enterprises were already based in that area, or how keen the local council was to support firms. By doing just this, they were failing to offer actual updated information and services that could be of some use to the local businesses. Another example was providing descriptions about the cultural heritage and history of art of a certain city, without informing about what was actually going on in terms of cultural events, let alone the chance of, for instance, purchasing tickets via the Internet. Needles to say that citizens are likely to know something already about the monuments they see every day, and that they would probably ignore an urban information system that does not provide something more useful to them. Another aspect that has to be highlighted, as it is not clearly shown by the chart, is that if the provision of information was generally scarce, interactive services, like the ability of performing money transactions to purchase goods or pay local taxes, for instance, were completely absent. This means that a vast majority of the initiatives were even failing to address those that could easily be envisaged as the most basic information needs of citizens, such as public transport timetables or information on local events, prioritising a brochure-oriented type of information. Again, this shows the dualism between global interests and local ones, and the fact that the former prevailed in many of the initiatives.

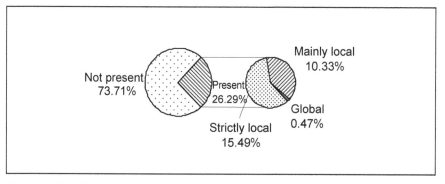

Figure 4.21   Web cities and provision of web sites to local organisations, related to local targets

Figure 4.21 displays the same results that were shown by Figure 4.8 on the provision of web sites to organisations, by the initiatives, but it specifies whether these organisations and firms appending information to the sites were locally based (or related) or not. As the chart shows, when it came to participation from external providers of information, nearly all of them ended up being local, small and medium enterprises and non-profit organisations. Indeed, over half of the initiatives that were offering web spaces to third parties had opted for these spaces to be available only to locally based organisations. But even in those cases where no special rules were specified, the overwhelming majority of the participants ended up being local anyway. This is expressed by the 10.33% of 'mainly local' information providers shown in the chart. These results demonstrate that whenever facilities for self-representation were offered by digital cities, these chances were being taken by local actors willing to participate in the initiatives or just to advertise themselves to local audiences. Quite possibly, the big 'global' subjects were not in need of acquiring scattered web spaces, and were considering, in any case, sponsorship of the initiatives. On the other hand this could also imply a lack of confidence from big corporate companies with respect to the alleged local 'penetration' and effectiveness of this kind of web sites. It seemed that whilst the small and local – including local authorities themselves – were struggling to acquire global visibility, those who already had it were sceptical or not very interested in the alleged potential of digital cities for addressing local markets and communities.

Putting it all together

What kind of broad empirical picture was offered to us by putting together the answers to the research questions addressed by this survey? What was the scope and potential of the digital city phenomenon in Europe at the beginning of 1997? Were web-based initiatives going to establish some type of extension to the public sphere of our cities?

On the one hand we have already answered these questions by constructing and analysing a typology of the initiatives, derived from a scoring mechanism related to the single aspects of the analysed sites. The results of this typological exercise have been described at the beginning of this chapter.

However, it is useful to briefly summarise the results organised by four main themes:

- The 'owners' of the initiatives and the contexts in which they were being developed.
- The efforts made to establish a public, actively accessible cyberspace and encourage debate and exchange.
- The relationship between the initiatives and local economies.
- The relationship between the initiatives and local citizens and communities (i.e. how grounded they were).

These themes were central for the whole research, and are present in the following analysis of the two in-depth case studies. These conclusions can therefore link the survey very well to the next bit of the research and indeed of this book.

Ownership and socio-political and economic contexts

We have seen that the 'digital city' phenomenon was developing in Europe in a way that highlighted a weaker divide between Southern and Northern European countries than could have been expected. Some southern countries like Italy were indeed at the leading edge of the developments in terms of sophistication of the initiatives, together with northern countries like the Netherlands. Rather than contrasts in quantity and complexity of the initiatives, the main differences between countries were to be seen in ownership and agency. For example, whilst the public sector in Italy was driving the most complex and interactive experiments of virtual cities, the opposite tended to happen in countries where local authorities had a less primary and central role in urban governance. In these countries, like the UK or the Netherlands, sites run by local authorities alone tended to be mono-directional information services tackling a very limited range of urban information and communication problems. Many possible functions and benefits of digital cities were left to the initiative of private (often non-profit) organisations and sometimes joint-ventures between public and private, where the public sector did not have however a hegemonic role. Within this scenario, the low number of initiatives in France constituted an interesting anomaly that could possibly have a 'technological' explanation, as the presence of an earlier very successful information system like Minitel could have generated a lack of interest towards Internet based technologies in the urban context.

These observations suggest that the socio-political and economic settings of each country, and the consequent different relations between public and private sector, had a strong influence in the shaping of the initiatives. Despite the establishment in Europe of trans-national bodies like 'Telecities' and 'European Digital Cities' setting the aims of the movement and encouraging initiatives to exchange ideas and experiences, it was very hard if not impossible to define a 'model' European virtual city and its dynamics of development.

The establishment of public, participative cyberspaces

We have seen that the level of complexity of the majority of the initiatives, especially for what concerned the provision of bi-directional communication facilities, was very poor. Very few sites ended up even addressing the issue of providing debate facilities and facilitating communication and exchange among citizens and between citizens and local authorities. Very few initiatives offered free of charge web spaces to individuals and (at least) non-profit organisations, to allow self-representation within the city. Therefore, while we can speak of a widespread phenomenon of city-related web sites, we cannot do the same as far as the concept of public cyberspace is concerned. Most initiatives ended up being mono-directional billboards or brochures, not spaces where interaction was possible. Very few examples could actually have the potential of enhancing the public sphere of their host cities and towns, by adding a parallel electronic dimension to it and encouraging public discourse.

The relation between digital cities and local economies

There is no doubt that marketing was one of the main elements of virtual cities. However, it seemed that in the overall majority of cases the main marketing activity performed by the sites was that of promoting the city to potential investors and tourists. On the other hand, local firms and enterprises were seldom encouraged to take advantage of the electronic space of the digital city in any way.

The information provided that could be useful to businesses was poor or non-existent in most cases, and few initiatives were actually offering web spaces to firms and organisations. This somehow represented a contradiction. In fact, cities were being marketed to investors as ideal places to establish an activity, while at the same time the web sites were showing that the actual provision of hi-tech services and information to firms already operating in the area was scarce to say the least.

The generally low sophistication of most initiatives was probably the reason why they tended not to attract advertising and representation of major global firms and businesses, which possibly did not see the virtual city web sites as actually able to 'penetrate' effectively their local communities. It should also be noted that in some cases virtual city initiatives had to compete with other 'virtual business parks' already established on the Internet and usually better organised. In general it could be said that in those few cases where facilities for businesses were provided, a few local firms and organisations took the chance of being in the virtual cities, mainly with the purpose of acquiring a 'global' visibility for free or at a very cheap price. But the overall impression was that virtual cities did not look 'good' enough to the majority of businesses, as their ability of actually creating a local market by addressing local audiences was doubted.

The relation between the initiatives and the real cities and communities

The previous observations about digital cities and businesses, and digital cities and public space, already suggest that the majority of the initiatives had failed to develop a very strong relation with their physical counterparts and to somehow work towards the needs of the communities dwelling them. This impression is reinforced when we consider that even from the point of view of simple information provision, let alone bi-directional communication, we have seen that most initiatives had the depth of a brochure, rather than that of a complex civic information system. The phenomenon of the tension in city government between the interests of the local people and businesses and the need to market the city to investors coming from elsewhere was mirrored in virtual cities. web sites were not exempt from this kind of problem. Indeed, very often this was the main reason for their existence, as the potential of the Internet was being exploited to gain global visibility rather than take care of local citizens and firms. Moreover, although local languages were prevalent, no resident immigrants' idioms were adopted, whilst English was widely used to address foreign audiences.

Conclusions

As we have seen, the results of the survey work indicate that the digital city phenomenon in Europe was at the beginning of 1997 very much in its early days.

The widely practised exploitation of the urban metaphor in the construction of web sites did not mean in most cases that a cyber-place was being established to enhance and co-operate with the real physical urban space. Indeed, very often, supposed digital cities ended up being nothing more than electronic brochures. However, a few initiatives tended to stand out of the less interesting mass of city-related sites. These were classified in the survey as the 'holistic', 'embedded digital cities'. Embedded digital cities tended to be complex, providing virtual spaces for debate and self-representation as well as relatively deep and extensive information geared above all at the citizens. This was the type of initiative that could be most relevant to the purpose of this research, and it was natural, after having determined where the interesting examples of web-based digital cities could be found in Europe, to examine them better. Once having performed the general analysis allowed by the survey, some more in-depth investigation in a limited number of cases was necessary to acquire more knowledge on the processes of construction of digital city initiatives and on their characteristic beyond what could be observed through web-browsing software. Two embedded digital cities were chosen as case studies, Bologna Iperbole in Italy and Digital City Bristol in the UK. The next two chapters analyse these two case studies.

Chapter 5

History of a Civic Network: Iperbole in Bologna

Introduction

The analysis of the survey proved to be extremely helpful in determining the most interesting and advanced examples of digital cities and thence suggesting a group of possible case studies.

Within this group, two initiatives were selected for comparison: Iperbole in Bologna, and Digital City Bristol. The observation of the web sites related to these initiatives, carried out for the survey, had already provided some basic knowledge of these two projects. It was evident that Bologna and Bristol were two of the most developed and interesting digital cities in Europe. Iperbole Bologna proved this further by winning the so-called 'Bangemann Challenge' award for public administrations in early 1997, and it was one of the most prominent members of the 'Telecities' group, promoted by the EU.

A series of pragmatic factors had also an influence over the choice of the case studies. The fact that the researcher is Italian suggested that a case study in Italy was particularly viable, both technically and economically. An Anglo-Italian comparison was also quite desirable. Analysing the development of digital cities in two quite different contexts would be extremely interesting and valuable to give the research a cross-European dimension that would not be restricted to a comparison between similar northern-European situations.

Also, the availability of good contacts and links with key-persons involved in the shaping of the two projects was a crucial factor for the choice of the case studies. As Yin argues:

> Key informants are often critical to the success of a case study. Such persons not only provide the case study investigator with insights into a matter but also can suggest sources of corroboratory evidence – and initiate the access to such sources (Yin, 1994, p.84).

Both in Bologna and Bristol the presence of a known person who happened to be at the very centre of the initiative was extremely beneficial to the conduct of the case studies. Through these informants it was possible to select the sources of information – mainly interviewees – and optimise the whole process without having to conduct redundant exploration. In Bologna, the good relationship established with the Iperbole project manager was vital in gaining access to interviewees such as the city manager, who otherwise would have been extremely difficult to approach. A similar thing happened for Bristol, where the previously acquired knowledge of a key researcher in Hewlett Packard Laboratories opened up opportunities to be introduced to prospective interviewees both in Bristol City Council and the University of West of England.

This chapter is based on the field research on Iperbole that was carried out between the end of 1996 and the first half of 1997, visiting Bologna, collecting documents and interviewing the key actors involved, and on successive observations of how the project was evolving in the following years. The next pages aim at constructing a critical history of this initiative, by looking at how it was shaped and by whom. To offer a framework to the reader, as well as addressing the main questions of this book in an effective way, the narrative is divided into sections. We first deal with the initial phase of life of the project, who promoted it, how it was described and envisaged, and who was 'enrolled' to become an involved actor in the design and shaping of the initiative. But at the time the interviews were carried out a kind of 'second phase' of development, with some remarkable differences from the initial agenda, was underway, as the whole project was being questioned and rethought. The last section is dedicated to the analysis of how the perception and interpretation of Iperbole was changing, and how it was evolving towards a more mature and stabilised part of urban technology. Between the first and last sections, three other sub-chapters deal with specific themes such as the construction of electronic public spaces and the relation with the real city, the relation of the initiative with local economic development, and the 'construction' of users and the strategies employed to facilitate access.

Inevitably, there can never be a clear divide between these themes, and it is a natural thing that most of the data collected relate to more than just one topic and answer more than one question.

Iperbole

In January 1997, a series of European awards, collectively known as 'The Bangemann Challenge', were given to projects and initiatives that were recognised as making a particularly positive contribution to the shaping of the European Information Society. The Italian city of Bologna was amongst the winners with an Internet site that Bologna city council refers to as a 'rete civica' (civic network), but that could well be classified as an 'embedded digital city'. The name of the Bologna site was – and still is – Iperbole, an acronym for 'Internet PER BOlogna e L'Emilia Romagna' (Internet for Bologna and Emilia Romagna), and at the time of the award it had already existed for about three years. Because of its successes and the reputation that it gained, Iperbole became one of the most frequently referenced examples, together with Amsterdam's De Digitale Stad, of advanced digital cities in Europe.

The context

Bologna is a major Italian city geographically located in a position that makes it a natural node between Central Italy and Northern Italy. It can be considered a medium-sized European city with population slightly less than 400,000 and a quite dense hinterland, if we consider that the whole of the province had in 1997 over 900,000 inhabitants (data from Comune di Bologna and Istat-Census).

Traditionally, the whole region of which Bologna is the centre, called Emilia Romagna, is one of the richest regions of Italy. The province of Bologna is also a low unemployment area, with a rate of 4.6% in 1998, dropped from 5.3% in the previous year, which was much lower than the national average of 12.3%. Another important characteristic of Bologna is the presence of one of the oldest and most famous universities in Europe and the World, which still attracts a large number of students, about 100,000 people who are a 'town within the town' (Director of Communication, 1997).

Bologna and the whole region of Emila Romagna have gained a reputation in Italy and abroad for their efficient and often innovative approach to public administration. Since the post-war period, the several 'comuni' of Emilia have had an uninterrupted sequence of left-wing oriented governments, largely dominated by the Italian Communist Party and, after its re-forming and re-naming, by the so-called Left-wing Democrats (Democratici di Sinistra). Only in the local elections of 1999 a change of government took place, with a centre-right coalition mayor taking over. These local governments have worked throughout the years to increase the efficiency of public administration, while often undertaking initiatives seen as progressive and socially inclusive. For instance, Bologna offered free bus services during the 1970s, and had many initiatives like the 'centri anziani', neighbourhood centres for elderly people, that are very popular as meeting and activity places. These centres are still extremely successful and regarded as examples of best practice. Probably due to the general political orientation of much of the population, Bologna and Emilia Romagna are areas where co-operatives and non-profit enterprises are particularly numerous as well as strong. This makes initiatives that are run to be inclusive and participative particularly feasible and able to rely on a large number of volunteer workers, as well as funding from local authorities.

All of these factors, already present in Bologna, could be recognised as fertile ground allowing the development of an initiative like Iperbole, aiming to exploit the possibilities offered by the emergence of the 'information society' for the benefit of the general public. Also, the presence of a strong public administration together with a prestigious university – and the fact itself that some influential people operated in both the environments – allowed important synergies and consensus to be created around the initiative.

Citycard and beyond: the early days of Iperbole

Iperbole is an evolution from another project, called 'Citycard', which started in 1993 as an EC-funded project under the ESPRIT programme. Citycard was project number EP8123, and had been initially conceived as an experiment of teledemocracy and public participation in decision-making in cities. The project manager was a relatively small software house based in Bologna, called Omega Generation and specialised in natural language analysis and artificial intelligence. Three European countries – Italy, Spain and the UK – were involved in the project, all of them with a technological partner working together with a public administration. In Italy the partners were Omega Generation and Bologna City Council, in Spain the municipality of Barcelona was working with Lagotex LDA, and in the UK Wansbeck District Council was liaising with Mari Computer Systems, an IT firm based in Gateshead, Tyne and Wear.

The name of the original project itself gives a clue to what the experiment had to be about, at least in the initial intentions of the promoters. Citycard drew its inspiration from a pre-existing initiative in Bologna, that was working well (and still is at the time of writing) called 'CUP card'. This is a magnetic card that every resident in Bologna receives free of charge from the municipality, and that can be used at 30-40 appropriate manned 'access points' scattered around the city to book medical services in Bologna. The initial idea for Citycard came from the director of Omega Generation, also a lecturer at Bologna University, who decided to write a proposal for an EC funded project, envisaging the possible use of 'civic' smart cards to survey citizens through a network of kiosks. Such an objective, however, had the potential to be extremely controversial. Some partners in the project, Wansbeck above all, according to some of the interviewees (project manager, 1997; and director of Omega, 1997) very soon rejected the feature of the civic 'referenda' as unacceptable and a possible threat for personal privacy of the end-users. The director of Omega himself seemed not too convinced of his own initial proposal. He was aware that, beyond the hype of the teledemocracy issue – extremely useful apparently to gain approval and funding of the project by the EC – lay the problem of the un-representative groups that would have used the system. During the interview, he minimised the importance of the initial idea about referenda, claiming that

> these were not meant to be formal referenda. They were ways to gather opinions, for which the fact that the respondents could be representative of the whole population was not so important. We were aware of who was able to respond and who was not, and we aimed at something more qualitative than quantitative (director of Omega, 1997).

This was a problem that was going to characterise the participative features of Iperbole as well. If the results of surveys, consultations or just debate among citizens were not to be considered reliable and representative, then did this kind of exercises make sense in the first place? As was the case for Citycard's referenda, which were soon shelved by the partners of the project, a very similar problem was going to happen with Iperbole's discussion areas, as we will see further on in this chapter.

But in 1993, and even more in 1994 – the year in which Iperbole would take over Citycard – the emphasis on the potential of IT for enhancing the urban democratic processes was far stronger than any concern for actual effectiveness of communicative tools of this kind. Citycard was an initiative undertaken by the best-regarded left wing local administration within the whole country, and it was something involving high technologies. Two senior figures with a strong vision for IT initiatives were among the main actors determining the birth of Citycard and its initial orientation towards public participation: the former Council secretary to Innovation and Renzo Imbeni. The secretary to Innovation, as we will see better in the next section, was a local decision-maker, member of local government who teamed up with the director of Omega in launching the idea. He remained in a central position within Iperbole as well, at least in its first phase of development. Imbeni had been an extremely successful and popular mayor of Bologna. He had recently become MEP with PDS, and his support within the EC was fundamental for the success of the Citycard funding bid of 2.5 billion Liras (Visani, 1993). The local as well as national press, mainly left-wing oriented – and very popular – national newspapers like La Repubblica and L'Unità started getting enthusiastic about the potential of the phenomenon. Most – if not all – of the articles written about the embryonic Citycard were extremely positive and sometimes hyperbolic towards the initiative and what it represented. These articles showed how strong was the emphasis towards the political and participative aspects of the experiment, that were seen as its primary purpose. Stefano Rodotà, a prominent and very respected politician of the PDS, interviewed by La Repubblica argued that 'Good use of these technologies could allow an increased involvement of the citizens in decision-making and design processes' (Rodotà, quoted in Marcoaldi, 1993, p.31). A month later, Valerio Varesi, another journalist of La Repubblica, would reinforce this discourse by talking of Citycard in terms of 'direct democracy... a continuous Swiss-style referendum'. He was so enthusiastic that he went so far as to compare the potential social benefits of this network of civic terminals with those of complex benefit systems such as the 'atelier de la loi' in France or Legal Aid in the UK (Varesi, 1993).

This climate of easy enthusiasm for IT initiatives that involved local democracy, together with the broader – and very agitated – political climate of the 1994 general election, were the ideal ground in Bologna for getting consensus for the development of a long-term project like the Iperbole civic network. The new project in fact – though not formally – was going to replace and re-define Citycard, as its initial propositions for direct democracy had quickly proven unfeasible but, above all, undesirable.

Launching Iperbole: actors and visions

In early 1994 the Iperbole project for a civic network was launched in Bologna. As we have already briefly mentioned, the context offered many reasons to several actors to get involved and/or support this initiative, so that it could actually take off. The fact that Citycard had already been funded by the European Community was certainly a major facilitating event.

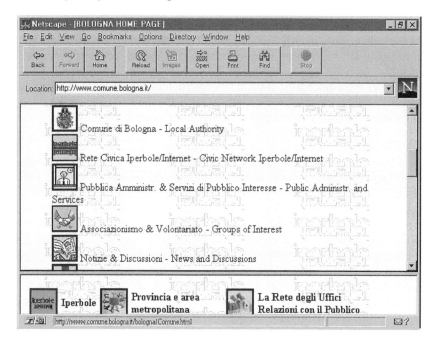

Figure 5.1 The Iperbole web site in its early days

Some crucial actors had formed the original project-managing group in Bologna. The Council secretary for Innovation as a member of local government, The Omega Generation director and his firm, and the Ufficio Comunicazioni coi Cittadini (Office for Communication with Citizens) of Bologna City Council. Each of these core members of the project had of course some very good reasons to promote the initiative and work towards its success. The following paragraphs will explore why Iperbole was part of these actors' strategies.

The vision from the (former) secretary for Innovation

The former Council secretary for Innovation is professor of Philosophy at Bologna university, a well-known intellectual who has been linked to a circle of friends all previously belonging to extreme left-wing political groups. Among his best friends and possibly advisors were left-wing writers, intellectuals relatively well known nationally and local cyber-gurus, writing on alternative lifestyles, cyberpunk culture, and innovative forms of democracy.

The former Council secretary for Innovation certainly had been the main source of inspiration for the Iperbole project in its early days, as well as its most prominent promoter. Sensitive to the relevance for local governance that the emergence of the information society could have, he envisaged mainly two reasons why the Council should have got directly, quickly and heavily involved in setting up a public Internet system for the city.

The first reason was tackling complexity and fragmentation in urban governance. The challenge here was to create an environment that could empower single citizens but above all organised groups of citizens, to self-represent and participate in the decision-making processes within the city. Still in 1997, when the interview took place and he was in a much more marginal position within the project, if at all involved, he was keen to stress that

> My position stems from the fact that in a complex society it is extremely difficult to know how the system works, being able to take decisions (...) Actually, even in Bologna the urban social system is far too complex for the administration to be able to satisfy people's needs. There is a deficit of resources as well as knowledge. Is it then possible to run and administrate the city effectively? My answer is no, it is not possible unless we manage to cope with complexity, broadening our knowledge base and the number of those involved in decision-making. I am not referring to direct democracy, but rather to a broadening of representation. We increasingly need forms of self-representation, and the net can allow people to organise themselves beyond traditional structures like unions or certain associations. People can speak beyond traditional representation forms, and beyond differences in wealth and social status.

The second main reason for creating Iperbole was the need for universal service. The secretary for Innovation actually saw telematics networks and the Internet as another utility – and a very important one indeed – that needed public intervention to assure that public access to ICTs would be genuinely public. The project was then seen as an instrument to promote the use of the Internet with people from all social classes in Bologna. This was of course meant to be a necessary condition for the accomplishment of the first objective of enhancing urban governance. Interestingly enough, the secretary for Innovation would not consider solutions in which public and private sectors would work in partnership. Although as we have already noted Iperbole was set up by several actors, the Council would retain a leading role and be, in fact, the only policy-maker in the shaping of Iperbole. Especially in its early days, the influence of the private sector was regarded as undesirable, as it would have introduced market-generated divides between information 'haves' and 'have nots':

> We thought that the function of promoting the use of telematics had to be reserved to the public sector rather than to the market. The latter would have obeyed the rules of wealth and cultural level, and the process would have become exclusive rather than inclusive. A market-driven promotion of mass telematics would also have been much slower.

This vision will generate one of the strongest policy initiatives of the first years of Iperbole, the offer of free Internet connections to all citizens who applied for it, as we will see later in this chapter.

The Office for Communication with the Citizens

What the Council secretary for Innovation had envisaged needed to be put in practice and given a precise role within the municipal machine. A senior officer from the Servizi di Comunicazione coi Cittadini (Office for Communication with Citizens) became – and still is – the manager of the whole project, initially under the supervision of the director of the Office. The project manager saw her office's involvement as an obvious thing, claiming that

> The Servizi di Comunicazione were the ideal place to accomplish the project, as we had already a strategically very similar series of activities. For us implementing Iperbole/Citycard was just adding a new tool to do what we were already doing.

Indeed, the role of the project manager soon became central to the whole initiatve. Having to organise and run the civic network, liasing at the same time with the policy makers, the developers, and the potential users of the system gave the project manager all-round knowledge of Iperbole and the processes that were shaping it. She was actually at the helm. Therefore, for the carrying out of this case study, she was the overall key person to rely on, and the source and facilitator of all other contacts with prospective interviewees.

The project manager and the director of Communication entered the project as the operative unit that took it on board and managed it within the Municipality. Though the general vision from The secretary for Innovation was in principle shared, especially by the project manager, because of their position they had a conception of the project that was much more centred on the role of the Council and its relationship with the citizens of Bologna. The project manager for instance described Iperbole's aims as a way to enhance the relationship between citizens and public administration, but stressed the project's value as 'A tool for administrative transparency and the internal efficiency of the Council'. Although the emphasis here was quite obviously focused on aims that were more practically and directly related to the way the Council worked, there was still a general belief that bi-directional communication and support of public discourse was going to have a central role in Iperbole.

Omega Generation

Omega Generation was a relatively small software house, directed by a businessman who is also a professor at the University of Bologna. As already noted, Omega was one of the original partners to begin the Citycard project accessing the European funds available in the ESPRIT programme. Obviously Omega had commercial interests that justified its participation in the project. They could be among the first to develop a public civic network in Europe, experimenting in the field several routines and ideas they had, as well as getting new ideas from the continuous feedback of their partners and the end-users.

This allowed Omega to successfully exploit this experience and market the software created for it as a package named 'MV Iperbole'. This package including web design of pages for civic networks and specialised routines for the management of bi-directional communications between citizens and public administration, was being marketed mainly towards other cities and towns in Italy. Several Italian local councils wanted to follow suit the example set by Bologna. Obviously, the very software solutions that were being implemented in Bologna and were being offered commercially by Omega Generation were certainly of great interest. Cities like Ravenna, Vicenza and San Marino, together with regions like Val D'Aosta became clients of Omega Generation embracing the Iperbole paradigm.

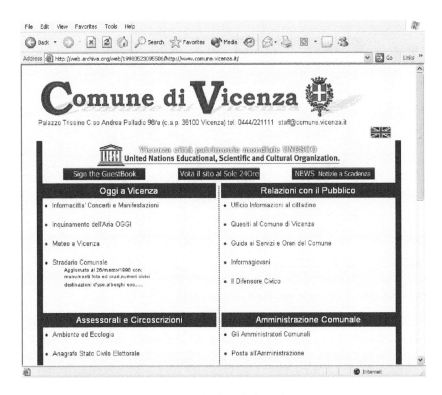

Figure 5.2 Vicenza's site based on the Iperbole software

What is of interest for this research is the fact that the expertise that Omega could offer was, again, centred on bi-directional communication and the electronic analysis of language. Omega were able, for instance, to set up a system by which citizens could email the city council on specific inquiries and problems without having to know whom to send the message to. The software would facilitate the process by comparing the contents of the email with a matrix of keywords and then direct the message to the right recipient. The director of Omega was also very much in favour, and able to provide technical assistance, of the establishment of discussion forums, using the well-known Internet 'newsgroups' technology, to encourage public dialogue in the digital city.

Figure 5.3 Ravenna's site based on the Iperbole software

As the project manager claimed in one of the interviews: 'In Bologna we found an extraordinary convergence of political and industrial objectives. The particular industrial objectives were going to suit very well the political and policy ones'.

CINECA, the University as Internet Service Provider

CINECA was the third piece of the jigsaw that Iperbole needed to operate. As the project management and the supply of information were coming from the council, the software development from Omega Generation, the provision of Internet connectivity and the relative technical assistance came from CINECA. This organisation is a public entity, being a consortium created by several universities in Italy to provide centralised high profile computing facilities and fast computing speed. CINECA is based in the immediate outskirts of Bologna and was seen as the ideal partner by the project management team in the Council. Several reasons were presented for this choice, that was going to be a controversial choice indeed, provoking an outraged reaction from several private Internet Service Providers (ISPs) in the Bologna area.

Towards the end of 1993 and the beginning of 1994 CINECA was in an expansion phase, opening up its services to non-scientific uses (Director of Omega, 1997; director of CINECA, 1997) and this was obviously presenting opportunities for collaboration with the Municipality. Another reason, on the technical side, for contacting CINECA was that in that very period, it had established a direct Internet connection with Paris (and from there to the rest of the World) that was considered quite fast and efficient (project manager, 1997).

As far as the consortium was concerned, as it was exploring ways of differentiating and expanding its activities, it saw joining the Iperbole team as an ideal chance to develop new synergies. The director of CINECA stated in his interview that 'It happened naturally that we found ourselves dealing with these people, and this was for us a stimulus to think that we were ready for this kind of step'.

CINECA entered as a non-profit partner as, according to its director, the Council was paying sums of money to them that were just supposed to cover the costs of material and human resources involved. He would however recognise two possible benefits that made it worth the effort for his organisation: in the short term it was enjoying a strong enhancement of its image. In the medium and long term the director of CINECA was not excluding the possibility for CINECA to become a successful ISP, able to compete in a local marketplace that it had significantly contributed to create.

The choice of CINECA as a partner, and its 'enrolment' in the network of actors involved directly in the shaping of Iperbole, however justifiable it could have been, reflects a prevailing attitude from the main decision-makers within the project. The former Council secretary for Innovation, the project manager and the Council were acknowledging a great divide between public and private initiatives and providers, and opting for a set up that could be as public as possible. The project manager would affirm that 'CINECA could provide a series of warranties because it was a public sector company, with a non-profit ethos rather than proper market objectives'. This perception of a clear and strong divide between public and private interests on the net, would inform most of the early management and policies related to the civic network in Bologna. This attitude will contribute to shaping a digital city in which the private, commercial sector will be almost totally absent. This, together with other impediments coming in as consequences of the general political climate, will limit the potential of Iperbole towards the support of public discourse considerably, as we are going to see later on in this chapter.

New media, transparency and elections: intepretative flexibility in the acceptance of Iperbole

Having described briefly who the main entrepreneurs were in the Iperbole project, and what their own visions were, it is important to get a perspective on how the initiative was described to gain support and acceptance within the current political climate. In other words, how was it possible for the secretary for Innovation, the project manager, the directors of Omega and CINECA to have the decision-makers in Bologna on their side?

Certainly some starting conditions were favourable to the acceptance of Iperbole. The fact that The secretary for Innovation, for instance, would play a double role as a local politician and mayor's advisor as well as the key vision-maker and creator of the project, was undoubtedly a facilitating factor. In his interview, he made jokes about the easiness of having Iperbole accepted by saying:

> One of the big advantages that we had when we proposed the project [to the City Council] was that nobody was really able to understand what we actually wanted to do, and this gave us the green light.

This might be true, of course, and indeed the weak reaction of the local political opposition to this project could be a sign of this. However, claiming that the reason for the Council to support Iperbole was just a matter of ignorance would be grossly reductive.

Two main ways of interpreting the initiative can be seen as playing a major role in convincing politicians as well as the press and the Bolognese public opinion to welcome Iperbole. The first is the construction of the image of cyberspace as the alternative, left-wing medium that could contrast the predominance of centre-right wing use of television. The second is the need for administrative transparency, being promoted by central government.

Cyberspace vs televison, the new frontier of the left wingers

As we saw before, Iperbole was born in 1994. This was a very special year for Italian politics. General elections were held in March, and this was the first time in which two major coalitions were going to face themselves. A change in the electoral law, after a national referendum held the previous year, had ruled out 'pure' proportional representation, and meant that parties had to form coalitions to be able to compete. Together with this, another major change had occurred: the phenomenon known in Italy as 'Tangentopoli', that is a chain of scandals and prosecutions for bribing major political characters, mainly belonging to the two ruling parties of Democrazia Cristiana (Christian Democrats) and the Partito Socialista (Socialist Party). Tangentopoli's main effect, in political terms, had been the total apparent annihilation of these two parties and the 'displacement' of a vast number of voters, who had ended up lacking their previous references. This situation generated another crucial event: the establishment of a new party called 'Forza Italia', placing itself in the centre-right area of Italian politics, trying to fill the gap left by the demise of DC and PSI. Forza Italia was created and led by Silvio Berlusconi, a well known media, housing and financial tycoon who had previously supported Bettino Craxi's Socialist Party. The elections of March 1994 were indeed won by the right-wing coalition led by Berlusconi, who had managed to offer a convincing alternative to many former Christian Democrats as well as Socialist voters. He was boldly presenting himself as the ultimate defender of freedom against the alleged risk of a Communist drift, as well as promoting a series of neo-liberal, Thatcherite policies.

The involvement of Berlusconi in Italian politics immediately provoked strong criticism from the left wing parties and, in particular, the former Communist Party now called Partito Democratico della Sinistra (Left-wing Democratic Party). The fact that Berlusconi, among his businesses, owned the three biggest private terrestrial television networks in Italy, was seen by his opposition as an extremely dangerous situation, in which the right-wing coalition could produce a phenomenal amount of propaganda and advertising for free. Indeed this had been, according to the PDS and its allies, the factor that had made the difference in the elections, together with the improper public use of surveys and statistics by Forza Italia, to justify its policies and programmes. Berlusconi was certainly surrounded by extremely good professionals who were able to exploit television and uni-directional communication. The use of this became more evident, fulfilling the fears of the left-wing opposition, when the Polo delle Libertà (Freedom Coalition), the coalition led by Berlusconi, got into power. Their win meant that, as the ruling government, Berlusconi and his allies could also exert power over the state-owned television network, RAI, by nominating its managers and news directors.

It is not within the scope of this book to investigate these events in-depth, or to construct a critical analysis of the use of Italian television in the 1990s. However, the context set up by these events clearly shows how the left-wing parties came to be extremely sensitive about the political use of the media, perceiving a situation in which the most powerful, uni-directional medium of national television was being misused to their disadvantage. According to left-wing leaders, voters were being cheated by Berlusconi's propaganda machine, and television was showing more than ever how unreliable and anti-democratic an instrument it could easily become. Within this context, computer mediated communication and the rise of the Internet could easily be seen as some kind of fascinating alternative to the power of television in Italy. Moreover, whilst the perception was that the battle for television had already been lost to Berlusconi, telematics were still a pretty unexplored and virgin territory. Finally, computer networking was something that had been developed locally in other countries, such as the 'FreeNet' experiments in the US, and that was going to be developed locally in Italy as well. Iperbole therefore came at the right moment, and could be interpreted in a very positive way by Bologna's mayor and its allies: an initiative by a progressive left-wing dominated local government that was going to show the way to innovative communication and democracy.

The Council secretary for Innovation (the municipality's 'assessore alle innovazioni istituzionali'), being the official advisor to the mayor on these matters, could therefore successfully promote an interpretation of Iperbole as the new frontier of the local exercise of democracy. It is also understandable why he and the rest of the Iperbole team put a very strong emphasis on the need for universal service identifying the Council as the only entity able to grant access to all. Within this interpretation of the initiative and the reality that was surrounding it, only a publicly-owned network could guarantee genuine and reliable communication, which television had already betrayed.

The major left-wing oriented newspapers would contribute to this vision of a 'good' use of new media by Bologna. *L'Unità*, the official and well respected national newspaper of the PDS wrote:

> The politicians have talked of electronic democracy (a EC funded project is being set up in Bologna that will allow experiment with telematics surveying and referenda) and of the importance of a democratic government of IT. A warning against widespread berlusconism (Visani, 1994).

La Repubblica, the second most widely read newspaper in Italy, made use of similar arguments to launch a frontal attack against Berlusconi. An article of 1994 (Smargiassi, 1994) used the news about the Bologna experiment to criticise the right-wing government every few lines. Another one drew an Orwellian scenario by stating that

> Against the risk of a Berlusconian Big Brother, the centre-left mayors put many little computers. From the political point of view this is called 'teledemocracy'. It is not a uni-dimensional communication, from the top to the people, but the ability to dialogue (La Repubblica, 1994).

In his interview, three years after Iperbole's birth, the director of Omega plainly acknowledged that the use of this paradigm was one of the main reasons for Iperbole's early success. He stated:

> In that particular moment [during the first Berlusconi's government] everybody was talking about information monopolies, and the paradigm of bi-directional communication was opposed to television. This gave Iperbole a political character that went well beyond our initial intentions. If the initiative was proposed now, it would be very unlikely to have the same effect.

The ever-present hyperbole about quick diffusion of new technologies was also reinforcing the paradigm of Iperbole as an innovative tool for democracy that was inevitably going to beat the arrogance of television-based communication. L'Unità had no doubts affirming that 'Within one or two years the way to be a citizen will be completely changed by ownership of computers and modems. These are now expensive devices, but tomorrow they will be like our homes' doorphones' (Curati, 1995).

Computers and bureaucrats: the challenge of 'administrative transparency'

Another important, different 'selling factor' for Iperbole stemmed from the need for more transparency in the way public administrations worked. As the director of Communication explained in his interview, the 1990s saw the beginning of a new concern for facilitating communications between public administration and the citizens. In a country like Italy, traditionally strangled by long and over-complicated bureaucratic processes, being able to make things easier and above all clearer for the citizens was – and indeed still is – perceived as an absolute priority for the renovation and regeneration of the country itself.

Computers started to be seen as the fundamental tool to transform bureaucracy and the relationships between administrations and people. Thanks to the hype dominating the debates about the potential of Information and Communication Technology, civic networking experiments such as Iperbole were seen as something that was going to re-define what being a citizens could mean. The power of the Internet was seen as something that was going to make it inevitable that the Council would change and improve, since using the civic network and ICTs in general was going to empower people and make them more aware, better informed, more critical, and more politically active. The director of Communication, talking about the difficulties of the bureaucratic machine to adapt to the speed of electronic communication, defined the problem as 'a complex organisation [the City Council] that now has to respond to an extremely exigent telematics elite'.

In particular in 1996, after the fall of Berlusconi's government and the appointment of Romano Prodi as the new PM, Franco Bassanini, a PDS member who had just become Ministro della Funzione Pubblica (Minister of Public Affairs), started issuing directives and laws to try and make bureaucracy faster and more accessible. Bassanini's team included, certainly not by chance, the Bolognese director of Communication. In the following years Bassanini and his team developed the idea of a national network for public administration, deriving the idea from the experimental local civic networks like Iperbole. This would give further momentum and motivation to people like the director of Communication, and part of the bureaucratic Bolognese machine, as himself noted in 1997:

> Sooner or later our network will 'come across' the Public Administration Unitary Network. This project is being taken forward at national level, and the Unitary Network will try and integrate itself with the pre-existing local networks. Those cities that will already have such initiatives will then have a clear advantage respect to those that have been waiting for directives and intervention from central government.

Already Iperbole was proving to be a successful project in the administrative arena. As stated in a report at the end of 1996:

> Iperbole/CityCard system has been selected as infrastructure for the 'Polyfunctional services in accessing the Public Administration', a national project promoted by the Ministry of Public Affairs to co-ordinate the activities of different parts of the Public Administration in making available updated information and services to the citizens through the Internet (Bellagamba and Guidi, 1996).

Bologna in the lead

As ever, Bologna was strongly promoting its image of innovative, creative local authority within Italy as well as the rest of Europe. Of course Iperbole was not just an initiative that was allegedly going to promote democracy, public participation and administrative transparency. It was above all the first to do so supported directly by a municipality. The most direct Italian 'competitor' of the Bolognese network was at that moment what was happening in Milan with the 'Rete Civica

Milanese' (RCM). But RCM was an experiment launched and run by the Computing Science department of Milan University, and initially it was not directly promoted by the City Council. In the rest of Europe other experiments had been going on, and in particular Amsterdam DDS was by far the most famous of all, but DDS itself had been funded with public money for a limited amount of time, and it had been managed by a private, although non-profit, company.

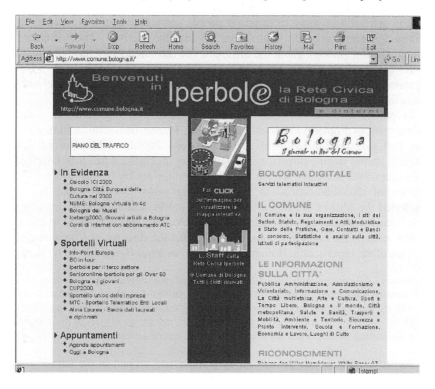

Figure 5.4 The Iperbole home page

Therefore, Bologna City Council was going to be regarded yet again as the most innovative and progressive municipality of Italy, and definitely one of the best of Europe as well. The involvement in CityCard had also resulted in Bologna joining the EC 'Telecities' programme as one of the founder cities. All of this had a political weight that went well beyond the local dimension. Bologna had been, since the end of the War, a stronghold of the left wing, and it was going to prove again how skilled and full of good ideas left wing administrators were. So, when the former mayor Renzo Imbeni resigned in 1993 to become a MEP, from his new position he kept facilitating the development of CityCard, shortly to become Iperbole. The new mayor Vitali immediately embraced the still embryonic initiative and embedded it in his policy programme (project manager, 1997). Vitali had to 'validate' himself relatively quickly as a worth successor of the extremely popular Imbeni, as in 1995 new local elections were going to be held, under an electoral system asking citizens to give direct indication of the person to become mayor.

Being first meant, among other things, that at least in the short and medium term the project would have been economically viable thanks to the funding coming from the European Commission. The director of Omega estimated that by 1997 CityCard/Iperbole had cost the Council about 1 billion liras, but that 70% of this money had come from the outside as EC funds.

Encountering a weak resistance

These characteristics of low-cost initiative that was going to win several international awards, within a general climate in which criticising telematics innovations would have looked like trying to stop an unstoppable progress, are probably reasons for an almost non-existent political opposition to the project. Criticising Iperbole would have been unpopular per se. All major interviewees provided similar information about weak opposition to Iperbole among local politicians and within Council sessions. The project manager said that only a few enquiries and criticisms from local politicians were generated, and that the answers provided by the Iperbole management team always ended up being satisfactory enough.

However, as we will see better in the next sections of this chapter, weak opposition came at a price that could be regarded as extremely high, as something directly jeopardising the spirit of the initiative itself: making Iperbole as neutral and free from conflict as it could be, in other words, influencing its contents and limiting political confrontation to a bare minimum.

Therefore, the only serious attacks to the initiative were going to come from local Internet providers, as we will see better in the section dealing with the end-users and the strategies to 'enrol' them.

Trying to construct public cyberspace

We have already seen how its promoters as well as a mostly friendly press described the potential of CityCard/Iperbole. Communication and participation were presented as the main benefits that would revolutionise the way local democracy was practised and achieved. When asked about Iperbole's potential to be a parallel, enhanced type of 'public space' for Bologna, the project manager, the Council secretary for Innovation and Mateuzzi all agreed that indeed this was the main purpose and merit of such an initiative. The director of Omega for instance spoke of 'electronic citizenry involving communication' adding that 'as we used to speak of rights to mobility, now we have to address these new rights'. The project manager defined Iperbole as 'much more than having another local newspaper: Iperbole is having a shared space'. Constant reference to the potential of Iperbole for supporting the enhancement of public discourse could be found in the literature as well. Stefano Rodotà, a well-respected left-wing politician had already praised

CityCard in 1993 by saying that 'good use of these technologies could allow an increased involvement of the citizens in decision-making and design processes' (Marcoaldi, 1993).

Both aspects of communication among citizens as well as communication between citizens and the local administration were highlighted as crucial components of a refreshed public participation to be achieved in the city. In 1995 the Council secretary for Innovation declared his commitment in the newspaper La Repubblica, by claiming that 'The Bolognesi connected to the network will be able to write directly to the mayor: a virtual mailbox will allow them to dialogue with mayor Vitali' (Sarti, 1995). Still in 1995, in the official documentation for the European Community about CityCard (although locally the project had already become Iperbole) a great emphasis was put on the function of public debate areas called discussion groups, or 'newsgroups', from the name this technology has in the wider Internet. The document argues that

> There is still more democratic promise to CityCard because it will be designed to enable citizens to discuss issues among themselves and to pass on proposed solutions to local problems. Discussion groups will even have the facility to produce edited documents for passing to the authorities (ISPO EC, 1995).

Indeed Iperbole had achieved much in terms of number of users and the amount of information provided on the network by the end of 1996. A growing number of citizens was getting connected, and the 7000 households of end of '96 were expected to double by the end of 1997 (project manager, 1997). We will see later in the chapter how these figures were not just met, but increased further in the following months and years. This unique result was mainly achieved thanks to a policy aiming to establish universal access by subsidising households' Internet connections, as we will see better in the section regarding the promotion of usage. At the same time all sorts of organisations based in Bologna were taking advantage of the free availability of space on the Council's World Wide Web server, becoming in fact information providers themselves, and contributing to increasing the complexity – and usefulness – of the civic network. At the beginning of 1997 these autonomous 'information providers' producing web pages were about 460. The now former Council secretary for Innovation noted how 'the capacity for producing information by the citizens should not be underestimated, if we think of comparing this with a more 'traditional' situation in which the Council would produce information and the citizens would just consume it'. In general the offer of web spaces to organisations was going to turn out as a very beneficial move. For instance, the spokesperson for an voluntary organisation assisting people suffering from epilepsy, claimed that

> we have built on the web a directory of social and health services for disabled people. Sometimes, by offering this publishing chance, the Council acquires information that it would be already supposed to have. So, through this initiative, we help each other.

On the discussion groups side, many areas for debate had been set up, well over 30 in 1997. These groups, initially open to Iperbole's members only, were then made available to people connected through other Internet providers, as it was noted that Bolognesi had the right to access the net through the provider they preferred, still being able to participate in the discussions. The groups' topics were extremely diverse, ranging from discussions about local political and planning-related issues to more light-hearted areas about jokes, sports or a sort of a virtual car-boot sale.

Despite these good starting points, though, it cannot be said that the establishment of an electronic space – or a series of spaces – supporting public discourse within the city was going to be an absolute success. Three main reasons can be identified, that were going to limit considerably the practical potential for Iperbole to be a powerful tool for enhancing the public sphere of Bologna. These are the limits imposed by the Council to freedom of speech and participation; the difficulties for the bureaucrats and personalities to adapt to a totally new way of dealing with the public; the endemic, technology-independent public participation problems that could not just be 'fixed' by introducing a new piece of technology.

Sanitising the public arena

The first and probably more crucial obstacle to the establishment of a fully participative cyber-space came from the way Iperbole was regulated from its very birth. We have already mentioned that the ability of the initiative to start and flourish enjoying a pretty weak political opposition had come at a price. Although this could have been necessary in the delicate political environment of 1994 and 1995 in Italy, the fact that Iperbole was regulated to be as neutral and conflict-free a space as possible, was inevitably going to limit its effectiveness. Iperbole was going to be a not too interesting arena from the point of view of political and social discourse.

For the project to be acceptable and shielded from political criticisms it had to rule out any form of propaganda and ideological statement. In practice it had to leave politics and activism out. It could of course have been conceived as a totally free arena. Leaving this space totally open for expression of whatever political position, however, would have contradicted the efforts that were being made at national level to limit Berlusconi's media's power in the public arena. Regulation was seen as paramount and, in absence of rules on Iperbole generated by the local politicians, the management team lead by the Council secretary for Innovation chose to put rigid limits on what could and could not be hosted by the civic network.

The management team of course was aware of the problems this was going to create. The discussion areas were left uncensored and non-moderated, but the ability to publish on web pages was heavily regulated. The project manager, talking about what was allowed in the 'public space' of Iperbole claimed that

> Information coming from the various users should have 'public' and 'generic' characteristics, and here lies the problem. Keeping ideology out is a very complicated matter, and it is a constant struggle trying to regulate what type of information can be

accepted from cultural associations, trade unions, parties and interest groups that have some kind of ideological or political character. These groups can generate problems.

Moreover, the users allowed to publish on Iperbole were only organised groups, while individuals could not obtain a web space. In 1996 The secretary for Innovation announced in the newspaper L'Unità that no political campaign was allowed within the civic network, defining any kind of declaration of support for candidates and political advertising as 'an improper use of the civic network' (Sarti, 1996). These activities, however, would have been perfectly normal in traditional public space.

This contradiction between the democratic participatory purposes of the project claimed by the secretary for Innovation and the others, and the level of political 'censorship' that was in fact put in practice, seemed to be a major problem of the whole initiative. Indeed, it could be one of the reasons that lead the secretary for Innovation to eventually resign from his advisory position, thus exiting the Iperbole scene. Although he would not speak directly about this, it apperas that he had tried to overcome this situation, by suggesting that all who wanted to have an 'ideological' space on the network should apply for this with the Municipal Council, thus forcing local politicians and decision-makers to regulate this aspect. He had also proposed allocating 2 Mb of web space to each applicant, declaring these sites as self-managed and allowing anything to be published on them that would not be either illegal or politically incorrect (project manager, 1997). But the Council did not bother facing the problem. The initial emphasis on the project as an electronic democracy initiative was starting to fade somehow, as we will see better in the section about the stabilisation of Iperbole, and probably the mayor and his colleagues preferred not to take risks.

The project manager, however, was very explicit in her interview about the non-sustainability of a censored public arena:

> We cannot keep the situation as it is, and it does not make any sense either. It is clear that we had to begin in this way, or it would have been hellish and unmanageable. But we have learnt a lot now, and we should really give ourselves a 'light' regulation to be offered to our 460 associations.

In general, the impossibility of publishing about sensitive issues certainly contributed to weaken Iperbole's initial role as a public space and an effective enhancement of Bologna's public sphere. Although debate areas were never censored, according to the project manager, it can be argued that it would not have been easy to get involved in open discussion of issues that could not be properly explained and publicised on the civic network.

Would you please reply? E-networks and old practice

One of Iperbole's main objectives was, as we have seen, to enhance communication between citizens and public administration. This was expected to bring two connected main benefits: the first was to increase the influence of citizens and groups over the decision-making processes within the city. The second was to improve administrative transparency, so that anybody could access quickly and clearly any relevant information, as well as being able to make inquiries of the right offices and people. E-mail messages were expected to be the main tool to achieve these goals, together with the 'advisory' role of the information to be found in public discussion areas.

Some very interesting and original efforts were made on the software development side to make communication between the public and the Council work. The most notable of these technological devices was probably the message routing system, which was very much appreciated by the users in a survey carried out in 1996 (Bellagamba and Guidi, 1996). Message routing meant that citizens trying to speak to the Council about some kind of problem were not required to know in advance who to speak with. Through an electronic 'form' on Iperbole any kind of query could be made 'to the Council'. The software, then, through matching the contents of the message with the contents of a keyword matrix, would find the right office or person to address the query to. This effort was the result of a six months preliminary study carried out on six European cities of the same size as Bologna. The director of Omega explained that

> the result of our analysis was that the primary need of citizens was not to have to understand the administration's complexity. This kind of difficulty was typically discouraging the citizen from talking to the Council.

He also added that

> the system analysing the content of messages was constantly being refined, though it had already reached an 85% accuracy.

So, citizens could easily write to the Council about all sorts of issues, engage in debates and send electronic messages directly to a series of selected personalities, such as the mayor, whose e-mail addresses were published on the Iperbole website. And indeed some of them started doing so, making the electronic arena potentially alive.

The main problem was notably on the other side. The director of Communication, as a deep connoisseur of the municipal machine, talked openly about the resistance to the initiative being experienced within the Council:

> The Italian experience in terms of communication with citizens stems from situations in which the public administration could either fail to dialogue at all or be tempted to make propaganda. Only in the 90s we have started thinking about transparent communication as a service to offer. And this implies a re-organisation of the way the administration works. Resistance is widespread but hidden. Nobody would openly disagree (...) Citizens write to the mayor and after a while start asking: 'why does he not reply?'.

The project manager highlighted the fact that changing the traditional slowness of Italian municipalities was going to be a long process:

> People are used to the rule that letters have to answered within 30 days, but replying to an e-mail message after a month obviously does not make sense. This organisation is a

slow and complex 'dinosaur' of 5000 people, and now it has to deal with an extremely exigent telematics elite.

However, a certain resistance within the system was quite evident. A user of Iperbole, coordinator of a non-profit cultural association dealing with theatre for communities and young people, confirmed that council officers she had to deal with had explicitly advised her to refrain from using email to send them documents and queries, and to use traditional letters instead. Despite some Internet training was being offered by the Iperbole team to Council officers, the level of refusal to take part in the initiative was still high. A senior planner contracted by the Council in a management position, claimed that

> Iperbole is used much more by the people connected from outside than by the Council workers. Nobody inside here uses the email, even secretaries cannot do it. I personally end up downloading electronic mail, making hardcopies of the messages and bringing them to the councillors involved. This is the situation we are in.

The management team was acting in two directions to try and cope with this difficulty: providing training and developing further software tools conceived to encourage, if not force, the officers to reply to e-mails. For instance, a piece of software was developed that would check whether messages had been replied within a few days, and, if not, would notify the Iperbole management team, as well as automatically sending messages to the faulty office, urging them to reply (project manager, 1997).

The director of Omega acknowledged another related problem. He explained that, together with the non-moderated discussion groups, some more 'high profile' forums had been meant to exist, in which important personalities, such as local politicians and intellectuals, would be involved. The director of Omega regretted that this part of the initiative had gone nowhere, as it could have been a crucial stimulus for public debate and participation. He gave the example of Umberto Eco as one of these Bolognese personalities who could have been convinced to be more active and write sometimes in these debate areas.

This general lack of feedback and true two-way communication and exchange of ideas was certainly another factor limiting the effectiveness of Iperbole as a space for public discourse. Even the discussion areas, which were supposed to be the main participative tool of the civic network, were going to lose part of their power, as we will see in the next section.

The failure of technological fixes to public participation

Because of its left-wing, partisan and anti-fascist legacy, as well as its tradition of post-War administrative efficiency, Bologna is considered by most Italians, and by the Bolognese people themselves, as a model of local democracy and participation. Indeed, Bologna and its region are the domain of Italian co-operatives, and citizens have a tradition of establishing associations of all kinds.

But this characteristic of many active citizens who take part in the social and political life of the city may have been taken for granted for too long. A senior planner had something to say about this in his interview:

> As in many other places, in Bologna the number of 'users' in increasing, while the number of 'citizens' is decreasing. I am not from here, but the Bolognesi claim that the city has changed a lot (...) Whenever we hold meetings in the neighbourhoods to involve local people in planning debate, we get greatly disappointed. Just about twenty people turn up, and they tend to be those elderly persons who were already used to participate twenty years ago. Here many take for granted that because Bologna is a 'democratic' city, then it is going to be forever as such. In the meantime, the recent referenda have been deserted.

It could be argued that a system like Iperbole had a potential to enhance public participation by providing, for instance, asyncronous communication – that is not having to be in the same place and at the same time to communicate – through the availability of e-mail and discussion groups. As contemporary lifestyles mean that it would not be easy for working, busy people to attend certain meetings, electronic communication could help. Also, the ability for the Council to publish, virtually in real-time, information about all sorts of issues, can be seen as another crucial advantage, and a facilitating factor, for an improved public participation.

However, there is no evidence that this was actually being achieved in Bologna. Indeed, there is evidence to show that Iperbole, three and more years after its establishment, had very little influence, if any at all, over the decision-making processes in the city.

The previous quotation from a senior planner's interview has a reference to referenda. These had been held on the 31st January 1997 and dealt with two important issues for the city. The first issue was about the privatisation of municipal pharmacies, while the second dealt with the design and location of the new railway station for Bologna. Iperbole was put in the centre of this event, as stated in the report on CityCard of the end of 1996:

> It is news of the past few days that the Municipality has decided to use the Iperbole/CityCard system as the preferred way to give citizens information on two big issues of political and administrative life of the city, that in few weeks will be arguments of popular referenda for all the inhabitants of Bologna: the decisions about the future management of the municipal chemists' shops and that on the new railway station designed by architect Bofill (Bellagamba and Guidi, 1996).

Iperbole was going to be the main information and promotion tool for the referenda, although voting was not going to be electronic, and people had to physically go to poll stations. Anyway, this attempt to introduce a sort of a technological 'fix' to the problems of participation did not work particularly well, or at least did not seem to improve the situation. As a senior planner was mentioning, the two referenda had to be invalidated because – as prescribed by Italian law – the minimum 50% of voters had not been reached. In fact only a disappointing 35.96% of people had decided to cast their vote (La Stampa, 1997).

Moreover, the lack of early success of Iperbole was going to generate a sort of a 'vicious circle' effect. Many officers within the Council were going to find it perfectly justifiable to ignore even more what was being said within Iperbole, as users were seen to be few, scarcely representative and, above all, they seemed not able to provide quality participation and more intent in e-mailing jokes on the system rather than to make interesting statements on city management. In fact the director of Omega noted that 'The messages in the newsgroups always come from the same few people' and a senior planner remarked that many administrators stated that, as Iperbole was an elite thing, they should not waste time and resources creating things for the Internet.

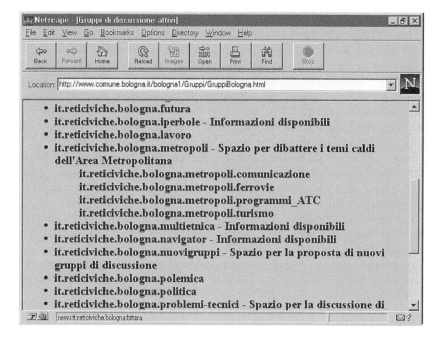

Figure 5.5 The newsgroups page in the early Iperbole

Even the director of Communication, a member of Iperbole's management team, admitted that, due to the poor contents of the discussion groups, these were rarely taken into account or used to improve the way the Council worked.

Indeed, observation carried out by the author two years later, in March 1999, confirmed that the discussion groups that were really active and utilised were the leisure and entertainment-oriented ones, while very little debate was going on about local politics, transport and general social and municipal issues. During a 12-days posting span,

the most successful discussion areas were 'jokes' with 110 messages, '2nd hand' with 107 messages, and 'computers' with 75 messages. On the other hand, 'multiethnic' had only 2 non-relevant messages, 'euro-citizens' had 3, and 'politics' had 8, just to mention a few (Aurigi, 2000, pp.39-40).

However, this does not mean that Iperbole could not have a role at all in enhancing public participation. The issue seems to be that, as suggested by a senior planner, the civic network should have been 'one of the tools' within a much more complex set of co-ordinated actions and initiatives towards revitalising a reducing participation. Expecting the civic network to represent a strategy per se, within a fragmented set of actions and intentions within the municipality, and with still a quite limited amount of capable Internet users, had been a flawed approach.

The local politician in charge of the office for Culture within the city, confirmed this assessment. As an important newcomer to the project towards the end of 1996, the Council Culture secretary acknowledged that Iperbole was not having a notable political and social relevance, as 'it does not have the possibility to affect the decision-making processes, and the process itself does not take its existence into account'. To improve the situation, he was envisaging an inclusive process that would enlarge the number of subjects related to civic problems, present on the network, so that a certain pressure could be built to influence decisions.

Integrating physical and virtual: Iperbole and the planning of Bologna

As a consequence of what we have just seen about the difficulties in establishing Iperbole as a public participation and public discourse environment, not much can be said about the civic network's influence over the planning of Bologna. Although some offices were starting to publish some transport-related information on the system, these early efforts tended to be limited to the uploading of existing, static text documents (senior planner, 1997).

A senior planner had been introduced by the project manager as possibly the most knowledgeable and passionate user of Iperbole among town planners within the Council, and he provided most of the little information available about the way that planners of the 'real' city were dealing with the electronic city. The planner interviewed claimed that most of the Planning department in Bologna was totally ignoring Iperbole and heavily criticised some of colleagues, arguing that not just the civic network, but any initiative geared towards participation was going to be seen as a nuisance. He said that 'very often planners here regard questions and queries from people as a useless obstacle to the planners' own knowledge and expertise' (senior planner, 1997). There were several ideas which he would have liked to develop within Iperbole, such as the establishment of an environmental Geographical Information System (GIS) and the management of discussion groups. He had also worked personally on the electronic information of the environmental impact of the new proposed railway station, and despite the poor results of the related referendum, he was still seeing a possible important role for Iperbole.

Although he was the first to acknowledge the limited interest of the general public in local policy-related issues, what he seemed to feel as the main problem was not the attitude of the citizens/users of the civic network. Several times he pointed his finger towards the institution itself and the fact that, as the Council Culture secretary himself had mentioned, more efforts had to be made to involve a large number of local actors, as well as providing more services and information for them. He claimed that within a perspective of combined efforts to enhance participation 'Iperbole would have been ideal, even if just one of the tools available' (senior planner, 1997). Still related to this general issue, the senior planner told of his own disappointing experience with a discussion group he had set up. Although seemingly very open and democratic, non-moderated discussion areas ended up being chaotic and dominated by a few protesters, he argued. Moderated and structured discussion would have been more useful, but this of course implied more work and resources for the moderation to be allocated (senior planner, 1997), and the impression was that these were lacking and the Council was not going to invest in them.

Figure 5.6 The 'virtual city' 3D interface option in Iperbole

In general, then, we can link these observations with what has been said about the establishment of public cyberspace and the related obstacles. The difficulties for the Council machine to adapt and validate its own public communication system, and the consequent lack of a true two-way exchange between citizens and decision-makers were heavily affecting the way the general public was going to 'adopt' the system.

Iperbole, local economy and the global-local tension

How was Iperbole's relation with Bologna's local economy and its development? Were local small and medium enterprises going to get a series of benefits from the civic network? Local economic development, although not central in the initial agenda, was nevertheless a purpose that had been considered for Iperbole, especially by mayor Walter Vitali in his 1994 campaign towards the local elections of '95 (Russo, 1994).

But electioneering claims apart, it has to be noted that, in respect to the main tele-democracy related aims, the relationship of the initiative with the economy had not been planned carefully enough. The main expectation and plan about using Iperbole to help enterprises in Bologna had initially involved the trade associations in the area. Trade associations were offered a presence on the civic network's website as well as free Internet and e-mail accounts. The project manager claimed that 'in our plans the trade associations had to be the promoting and 'multiplying' element to demonstrate what could be done on the Internet by small and medium enterprises, traders, craftsmen, builders and so on'.

At the same time, single enterprises were offered alternative Internet services and presence by CINECA on its 'Nettuno' network, but these were charged for, apart from a three-day free introduction course. Moreover, individual firms were literally kept out of the digital city, on the grounds that, as had happened with ideology and political propaganda, Iperbole had to be kept 'pure' and uncontaminated by profit-making initiatives (Director of Omega, 1997). Although he claimed that about 250 local firms had decided to apply for the CINECA account, it can be argued that the missing link between them and the digital city was in fact leaving them somehow 'invisible'. Certainly Iperbole was not able to facilitate the promotion of these firms' commercial activities, either locally or globally, as it could have done by hosting or at least 'linking' from the city site to their own sites.

As any kind of commercial activity was off-limits in Iperbole, non-profit organisation, hosted on the system, that wanted to use the network to sell services and create new opportunities for growth, could not do it. The coordinator of a community theatre non-profit organisation argued:

> We would like to widen the range of courses that we offer. Together with the photography, music and theatre courses we had thought to exploit our Iperbole account to offer Internet courses for children. A theatre asking to make their web site has also contacted us. The Internet allows us to be visible and to get known, but the Council should find a good justification to let us use Iperbole for business, as we are, after all, a non-profit association.

The Council Culture secretary also had plans and ideas about economic regeneration initiatives involving cultural enterprises, that required some changes in Iperbole's closure policies. He claimed:

We should invest in culture. I am not just talking about the computer literacy initiatives that are being carried out. I am talking about a new kind of entrepreneurship that would be able to exploit this civic network, cultural and multimedia entrepreneurs.

Another project, this time generated by the Council itself, and called 'Million', aimed at putting Bologna's hotels and tourist enterprises on the Internet and offering electronic booking services to those who were coming to the city. It was, according to the project manager, a difficult project to start, as many hotel owners were very diffident about the idea of putting data about themselves and their businesses on the Internet. However, even Million could not be hosted directly by Iperbole, and the site had to reside in Omega Generation's server, though a link from the Council pages existed in this case. As for the case of 'political' publishing, an official decision from the Council and its politicians about letting firms in was still awaited. The director of Omega was quite clear about this, saying that

> Projects such as Million are not to be considered part of Iperbole. These are value-added services involving economic transactions, and the fact that the civic network can host charged services has not been decided by any politician.

Indeed, other pressures not to open the civic network for business use were being put by the local Internet providers, and first of all by CINECA. As the director of CINECA candidly admitted:

> It is not up to ourselves to decide whether commercial enterprises should have a presence in the digital city, to access Iperbole's users. However, we and the other providers offer commercial services, and the Council should be careful and aware of this.

At the same time, the official plan to involve the trade associations as a catalyst was clearly failing, as response was limited:

> Participation of trade associations has been disappointing compared with what we expected. At the beginning we organised meetings with their representatives to encourage them, but eventually we stopped trying to involve them, after the first year of attempts (…). In my opinion the trade associations have interpreted the initiative as something that could deprive them of their mediation role, with the result of losing membership, or at least a certain kind of power over members. Their members at the moment have to go to the association to get useful information and services, while with an Internet tool they could get these directly (project manager, 1997).

A complex situation had therefore developed. The Council had adopted an approach that tended to recognise and involve just those bureaucratic trade associations that were supposed to support firms and enterprises. This approach implied that the role that the Council could play in local economic development was that of providing services and information to firms, but not that of creating markets, promoting these firms, or generating chances for business. Or, at least, this was not going to be done through the Internet initiative. The recent involvement of another actor, the city manager, as we will see later, was going to reinforce this vision of the Council as provider of enhanced bureaucratic services, though the city manager was approaching this from a totally different perspective than the original members of the management team. Paradoxically, whilst trade associations were to have promoted the use of the Internet with their members, the latter and their activities were not allowed on the civic network, not even through a series of links. They were expected to 'inhabit' a different electronic city from the one that hosted the citizens of Bologna. In real terms this could be compared to the Council intentionally banning shops from the streets and squares of Bologna, to gather them in an out-of-town shopping mall, with very little transport connections to it.

At the same time, even if the 'traditional' commercial sector was not very reactive, the Council Culture secretary's idea of facilitating the development of new cultural and IT-based entrepreneurship, just like the coordinator for the non-profit community theatre organisation would have liked to embark on, was being frustrated. The general ostracism of business and profit from the civic network, as well as the subtle opposition of the private Internet Service Providers, including CINECA, were preventing a full use of Iperbole for local economic development.

It can of course be sustained that this attitude could be beneficial because of the virtual absence of commodification of the information on the civic network, that was supposed to be there for all citizens and not to be 'bought' somehow. It is also true, though, that such a radical position was preventing new enterprises from finding a welcoming and cheap 'electronic roof' under which to develop, with potential loss of benefits, especially for the young population of Bologna.

'Constructing' the users: strategies to enliven Iperbole

All of the themes that have been dealt with so far are of course interrelated, and all depended on a crucial factor: the digital city needed 'digital' citizens or, in other words, those access related issues that have been mentioned in the literature review had to be dealt with. In the case of Iperbole this was even more crucial, as one of the main interpretations of the project was the need to allow access to the digital world to virtually any citizen:

> Access to this world [the Internet] is in my opinion a right of a new democracy, and therefore has to be granted by the public sector. Widening the use of the Internet beyond the initial elites was one of the aims of the initiative (Former Council secretary for Innovation, 1997).

In 1995 even the former and very popular mayor Renzo Imbeni had published an article in La Repubblica arguing how fundamental it was to tailor the information society around the needs of the disadvantaged citizens (Imbeni, 1995). This message was reinforced by the hyperbole present in many press articles at the time CityCard/Iperbole was born. The IT magazine *Zerouno*, for instance, celebrated the conception of the civic network by claiming: 'Bologna has opened the way of Internet for all' (Bassanetti, 1994).

But had it? It has to be said that much work and attention in the first three years of Iperbole had gone towards the target of encouraging and facilitating access to the civic system and the wider Internet to as many citizens as possible. Of the three main issues of support to public discourse, economic development, and universal access, the latter was surely the one that the management of Iperbole had tried to address most proactively. A steep growth of subscriptions to the service had started in the second half of 1996, when the 3,000 members of the first two years of life of Iperbole quickly became 7,000 by the end of the year and were forecast to double by the end of 1997 to 14,000 or even 15,000. These results were extremely good in terms of the ability of Iperbole to involve citizens. In mid-96 The Council secretary for Innovation had set the target of reaching 10,000 connections by the end of 1998 (La Repubblica, 1996), and soon after it appeared that this figure was to be reached well before that time. Actually, what had been forecast turned out to be slightly over-optimistic, and at the end of 1997 the individual users were almost 12,000. This was anyway still beating the target.

The results of Iperbole in terms of number of connections were therefore pretty encouraging. The director of Omega claimed that, considering those Bolognese people being online through other providers, as for instance those students who had university accounts, it could be estimated that towards the end of 1997 about 10% of the entire population of Bologna would have been using Iperbole and the Internet. The project manager also noted that the level of appreciation of the initiative by the users was very high, and that everybody felt the issue of facilitating access was very important. In fact, a certain amount of participation in the management and growth of the civic network was coming from groups of users. The project manager said that: 'several users are already collaborating actively, editing for instance the 'Frequently Asked Questions', and making themselves available to do some voluntary work to train new users'.

What strategies were employed in Bologna to achieve a number of subscribers and users which can be considered a very good result, if compared to other civic networking initiatives? And, above all, what problems were encountered and what could have been done more or better? As outlined in the literature review, three main issues about social access to telematics network have been identified and addressed. The first was about countering the negative effects of economic polarisation towards access, and something pretty bold for the times was done to deal with this, as described below. The second issue was about countering the literacy problems – especially computer literacy – to allow more people to use the several IT instruments available. The third issue, unfortunately not properly dealt with in Bologna, or at least not subject to any precise strategy and planning, was about the polarisation that comes from age and diversity in general, and the fact that people could be marginalised – or self-marginalised – by their inability to deal with close, homogeneous groups.

Free connections on offer

The Iperbole team, and Bologna City Council as the owner of the civic network, decided to attract users and tackle the problems coming from some of the costs involved in using the Internet by subsidising Internet connections and placing public access computers in the city.

Since the very start of the initiative, citizens were allowed to use the civic network for free, by dialling up the Council's modems. However, the number of World Wide Web sites that they could access was limited to Iperbole and a few other local information sources, such as Bologna University web site. In April 1996, though, it was decided to open up the whole Internet to Iperbole's subscribers, still keeping the access totally free of charge, apart of course for the costs of the telephone calls (Matteuzzi, 1996, p.15). In other words, what happened was that the Council started giving Internet accounts for free to any citizen who would apply for one. This was in line with The Council secretary for Innovation's thinking, as he had declared just a year before that 'With Iperbole, Bologna becomes a reference point, as it has decided that access to networks should not be affected by the ability to buy a service' (Il Resto del Carlino, 1995). However, this would not solve totally the problems related to the costs of network usage, nor did it address the costs of buying the necessary equipment. It was anyway certainly a way to encourage people who otherwise would have not spent their money to give the new medium a try. This has of course to be put in context of 1996, when Internet accounts in Italy could cost a minimum of about £150 per year and free deals with private providers were still a long way off.

As we have noted already, this was a very successful policy measure, as subscriptions increased dramatically and at the beginning of 1997 the waiting list to obtain an Iperbole account was as long as one and a half months (project manager, 1997). This situation generated what can be regarded as the most serious opposition to the project itself, which rather than coming from local politicians or pressure groups, came from the Internet industry. Straight after Iperbole's decision to offer free Internet connections, four local Internet Service Providers sued the City Council. They argued that CINECA had been chosen as a provider by the municipality without carrying out a proper bidding and selection process, and that the recent decision of giving away the Internet for free – although Iperbole's sessions had a 30 minutes time limit – was going to involve a loss of profits for the private operators. Together with these local and relatively small enterprises, two much bigger providers, Agorà, linked to the Radical Party and Berlusconi's right wing, and Technimedia had begun criticising the Bologna initiative for some time. Even if in 1994, when Iperbole was in its very infancy, it had immediately appeared to constitute a potential threat to businesses, and its philosophy of 'Internet for the masses' was under attack (Miccoli, 1994).

However, the people managing Iperbole dismissed the attacks as unrealistic. The director of Omega said that

> Our reply to them is supported by facts. The increase in the number of private Internet users [generated by Iperbole] makes the whole Internet related market improve rather

than decline. In fact Bologna, together with Florence, has the highest number of private local Internet providers.

The left wing oriented press welcomed the decision of the Regional Administrative Tribunal (TAR) to reject the complaint as a 'revolutionary decision' (L'Unità, 1996).

External opposition apart, the decision to grant free access – or rather subsidised access – to the Internet was going to be criticised and reviewed internally as well, as we will see further on in the chapter. However there is no doubt that this was the main reason behind the steep increase in Iperbole's memberships and a factor that surely facilitated usage.

But apart from their number, who exactly were the users, and who was Iperbole targeting?

Identifying the cyber-citizens

Membership of Iperbole was, of course, open to all citizens of Bologna, and given the objective of universal access of the initiative, no specific groups of people were excluded. However, we know already how exclusive the effects of social polarisation, as well as differences in education and age, can be when it comes to accessing and above all using effectively computer networks. Iperbole was no exception to this.

A report at the end of 1996 for the EC CityCard project, edited by Federica Bellagamba and Leda Guidi, provides a series of interesting figures about who the users of the system were. The total numbers were, as we have already mentioned, very encouraging: 6,200 individual citizens – and presumably their households – were surfing the Internet and using Iperbole together with 60 schools, 370 non-profit associations, 65 public administrations, 35 municipalities of Bologna's Metropolitan Area, and 125 municipal offices. A year later the individual subscriptions had risen to 11,720 and the non-profit associations to 490. But the most interesting data came from a user validation study carried out by the Iperbole team themselves. Two sets of questionnaires were produced, and answers to these obtained. The first set of responses came from people who had chosen to answer the questionnaire directly from the web, on their own initiative. The second set of responses came from a randomly selected sample of 200 Iperbole members who had been sent a hardcopy of the questionnaire. The differences between these two sets of responses give an insight on how polarised usage of the system still was, despite the efforts made to facilitate social inclusion by subsidising access.

Of the 200 home-sent questionnaires, only 40 were returned and used to compile statistics. Indeed, the home survey seemed to show some promising results in terms of reduction of polarisation in the use of telematics networks. Over 20% of respondents were women, and 8% of respondents were over 55 years of age. Also, the majority of people had relatively low education, as 52.5% were educated to 'medium school' level, which in Italy is the minimum legal requirement and stops at 14. In absolute terms a ratio of 80% men vs. 20% women obviously cannot be seen as a satisfactory gender balance in the use of new technologies.

Nevertheless, when put in the context of what these kind of initiatives were likely to achieve in the short term, it was indeed a good result.

However, the low response ratio of the home questionnaires – only 20% – and the fact that the 'home' respondents often tended to leave many questions about Iperbole's services and facilities unanswered, as if they had very little knowledge of the system, suggests that 'members' were not necessarily frequent 'users'.

The 'online' respondents, on the other hand, numbered 228. Their gender, age, and education distributions showed a very different picture. Among those who filled the online questionnaire, in fact, only 9% were women, and half of the respondents were aged between 25 and 34. Education levels were also much higher, as 50% of users had a 'secondary superior' school diploma – equivalent to 'A' levels in the UK – and 35% was educated at university degree level. The online questionnaires were filled in much more completely, showing that these people had a good knowledge of the civic network, as acknowledged by the authors of the report themselves:

> We can make some observations: people that answer the online questionnaire are, on average, younger and generally more interested than the people who sent back the compiled 'home questionnaire'. Indeed, they filled out the spaces for suggestions more frequently and also made various observations on the service, while the 'home users' often did not answer at all (...). Obviously, people that use the system and choose to fill the online questionnaire are more interested in the system itself (Bellagamba and Guidi, 1996).

The comparison of these two surveys, together with the data about the growth of Iperbole since mid 1996, suggest that the offer of free Internet accounts had indeed increased the number of households who were manifesting some kind on interest in the civic network, but that this initiative alone was certainly not enough to achieve the social access goals of Iperbole. It appears that a considerable number of household had joined the civic network because it was offered for free, but then had found themselves either not able, interested, or welcome to use it. The project manager was fully aware of this when she said that

> Data clearly reveal that a large part of our 7,000 users is 'sleeping'. This means that many of those who obtained the account have used it once, and because they were not able to understand what it really was have given up with it. We can see this by observing the discussion groups and the traffic in general. The fact that 80 modems are enough to let Iperbole work is meaningful, as the ratio of 85 people per modem is absolutely unrealistic.

It is also evident that the 'traditional' problems of polarisation were not easy to overcome, and Bologna's network was still dominated by young, well-educated males, just like the whole of the Internet.

Few resources to draw people in

The Iperbole management team, and the project manager in particular, were very aware of the limitations to their ability to involve and include users of all ages and backgrounds, and were reflecting upon methods to improve this situation, beyond the free Internet deal. Both the project manager and the director of Omega, for instance had the wish to involve those age groups who tended to be more marginalised, such as the elderly and children. The elderly especially could be seen as a very sensible target, considering that this group was already very numerous and growing because of an increasingly ageing population in Italy and Bologna as well:

> I have this dream: the elderly. 30% of Bolognese population is now over 60, and the elderly people of this area have very strong personalities and a proactive attitude. These people have set up the centres for the elderly, which are very well organised facilities, the allotments, the University for the elderly. I am putting in our agenda the need for an experiment in the centres for the elderly (project manager, 1997).

However, this was little more than a dream, as the project manager had defined it herself, and the director of Omega commented saying that 'the idea about involving the elderly is interesting, but we need a precise strategy that has not been determined yet. We cannot leave things to spontaneous initiatives'.

A senior planner was emphasising the need for complex initiatives that would go well beyond the offer of free accounts, arguing that

> Demand [for using the system] has to be created proactively. Giving free accounts is not enough and the general attitude is too deterministic. It seems for instance that setting up discussion groups creates the conditions for participating, while the inverse would be truer.

In general, several needs that had not yet been addressed properly could be recognised. On the one hand, the relatively young age of the civic network meant that its potential to inform was still low, and no 'real services, that have a practical effect' (project manager, 1997) were present in the system. The director of Omega, for instance, thought that the ability to offer electronic certification from the Council, through the civic network, would have become a strong attraction factor, drawing users in. This was seen as basically a matter of time, as some efforts were being made to develop interactive services and increase the number and quality of Council information on the site. Also, the recent involvement of new prominent actors within the Iperbole team, such as the city manager and the Council Culture secretary, was going to facilitate this.

But in general, resources to implement other vital measures to increase inclusiveness were pretty low. The installation of public access points, that is networked computers owned by the Council that could be used by citizens who could not afford to buy a PC, was well behind schedule. The director of Omega lamented that only 6 of these had been activated at public libraries and neighbourhood centres, in contrast to the target of 30 all over the city, and bluntly defined this as 'a failure, at the moment'.

Computer literacy courses and technical support were organised mainly for the Council's internal market, and low-cost Internet courses set up by private local organisations were encouraged and advertised. However, while the project manager was mildly optimistic about Iperbole's capacity for training and technical support, the interviewed users were less enthusiastic about the offer of support. The responsible of the voluntary organisation claimed that no Internet or web training had been offered to them, even if they would have liked to get it. He was satisfied of the technical support obtainable by telephone, though. A senior planner, as a user internal to the Council, complained about the courses for the employees by saying: 'It is not clear what the priorities are in holding the courses, but I am sure that the training is not working well'.

In general, it can be argued that while at the very beginning of the project, in its pioneering phase, the funds received from the European Community had been enough to sustain it, the post-96 boom of Iperbole was opening up a serious resource problem. As the director of CINECA suggested, 'the Council should get better equipped now. So far, in my opinion, they have underestimated the commitment that is needed to sustain such an initiative'. More money as well as people were needed to keep the momentum of the civic network's growth, otherwise the already present phenomenon of having many inactive members would increase further. And because more money was needed, the crucial issue of making the initiative sustainable in the long term was coming to the surface. The city manager was expressing this through a somehow paradoxical statement: 'If the number of users keeps growing so quickly, it will be a disaster in economic terms for the Council'. The need to 'rethink' Iperbole was being pushed hard, and this was going to change the overarching paradigm on which the initiative had been based, as we are going to see in the next section.

Towards stabilisation? Paradigm shifts in urban telematics

1997 was the year in which the conditions for a general reflection on the way Iperbole was developing were ripe. The Council Culture secretary was explicit about this by saying:

> We have to say that a phase has ended, that of Iperbole having a series of very limited purposes. Another, different phase is opening up, and this is not the civic network as we had defined it initially. I think we should organise some debate days with some possibly interested new actors to understand what Iperbole's phase 2 could be.

What was the state of Iperbole at the beginning in 1997? We have seen many aspects of it in detail in the previous sections, but it is worth summarising here to see why and how changes to the initiative could be envisaged.

Iperbole was a very successful project from the point of view of its reputation and consideration both nationally and internationally. It was indeed becoming a leader in Europe, and this was validated by the win of the 'Bangemann Challenge' award, that was going to be the first of a series of awards gained during the following years. This of course was opening up even more opportunities, as the municipality and Omega Generation were getting more and more invitations to conferences as well as requests to participate in European R&D projects (Director of Omega, 1997). The public image of Bologna City Council was boosted as the first municipality in Italy to be forward-looking and friendly to new Information and Communication Technologies. As a consequence of this, the name Iperbole had become a brand itself, as Omega had chosen to use it to market its civic networking software to other cities.

Relatively impressive figures in terms of membership had been developing especially since the launch of the free full-Internet deal in mid-96, but behind these numbers a certain apathy from the users was present, and levels of actual participation and active usage of the civic network were lower than expected. Cesare Maioli, a professor in the department of Maths of Bologna University, writing a paper on civic networks and Iperbole had noted that 'The project's vocation does not look well defined (...) the present bi-directional aspects cannot be appreciated' (Maioli, 1995). The people involved in the management of the initiative all perceived the need for richer contents in terms of information and interactive services, and some of them were also acknowledging that inclusiveness had to be boosted. More educational initiatives, as well as public access points in the city, were needed, and some parts of the population, such as the elderly and children, could have been involved through focused initiatives that had not taken place yet. But resources were running low. As Iperbole was growing, so had money and the number of people devoted to it, and the subsidising of connections by the Council was itself becoming very onerous.

These were the pre-conditions for re-thinking Iperbole in 1997 and the following years, a process that was going to be facilitated by the involvement of two new actors and their agendas: the city manager, and the Council Culture secretary, the 'assessore alla cultura' as this position is named in Italy.

The city manager and the Council as a 'machine'

Being the first city in Italy to have a city manager was another of Bologna's proud achievements, apparently. The recently appointed manager was an engineer by background, and the director of Communication defined him as the 'operative mayor', underlining the importance and prominence of his role in taking practical decisions on the Council's agenda.

The city manager did not seem to have a very detailed knowledge of the Iperbole project and its state of the art, admitting that he had never seen its web pages as he did not personally use computers in his job. Despite this, he had some extremely clear ideas about where he would like to see Iperbole moving to within the next few years. These ideas stemmed from his agenda for the distribution of services within the city, and his overall approach to the relationship between Council and citizens.

The project that the city manager was going to undertake was named 'Progetto Distribuzione Servizi' (Services Distribution Project), that quite ironically formed the acronym PDS, the same as the ruling party in Bologna. The project itself was not going to rely on the Iperbole/web platform only. The computerised points of CUP, scattered around the city, where citizens could book public health services and visits, was the existing base for PDS. Synergies were also being explored with local banks for the realisation of DIMMI, a sub-project within PDS that would allow citizens to use their electronic municipal ID and their cashcard to do money transactions with the Council at banks' ATMs. These transactions were initially going to include payments for Council taxes and parking fines.

The city manager got interested in Iperbole as he saw the opportunity of extending these services to the growing network of personal computers in private houses:

> The idea is to combine a series of 'pieces' (CUP etc) and get into every home. So that it is not important where the Council is, because it ends up being everywhere, a virtual council from which citizens can obtain services wherever they are.

But even more interestingly, he would go on saying:

> We are also re-thinking about the construction of CityCard. But a CityCard different from the 'Bonagian' one, something that could include not just the CUP, but also an electronic wallet that could be used to pay, for example, for the newspaper or for bus and train tickets. This is a project that banks are very interested in.

Basically, the city manager saw the primary role of the Council as that of an organisation producing urban services. He firmly believed that the practical consequences of offering good services to the people of Bologna had to take priority:

> My strategy aims to make the machine-city work and be more efficient. It does not tend to give answers about social cohesion and social functioning of the city. Those have to be dealt with by the mayor and the other politicians. However, it is obvious that a better working Council is a contribution to development and quality of life.

Within this interpretation, Iperbole as it had been until 1997, was seen as a marginally relevant attempt at managing Bologna, that could – if appropriately re-defined – provide a great potential for the distribution of services thanks to the relative success that it had had in terms of membership.

Cyber-services for clients vs cyber-places for citizens

PDS and the other ideas from the city manager were, in a way, something that Iperbole did need to validate itself as something really useful in everyday life, as well as being more attractive to the public. However, it was clear during the city manager's interview that a paradigm shift, rather than just an addition of services, was underpinned by his positions. As some of the quotations in the previous section underline, the city manager envisaged a radical re-definition of Iperbole that would put the accent on the fact that this was going to be a tool used by the Council to provide services to what he defined as 'the citizen/client' (city manager, 1997). It is worth noticing that the former Council secretary for Innovation – the 'father' of Iperbole – had been claiming a totally opposite conception of the relation council/citizen:

> When we meet colleagues from other European municipalities, and the English in particular, we willingly provoke them by emphasising the term 'citizens' in opposition to 'customer'. For us, 'citizen' is much more than 'customer' or 'client', because people must not be restricted to consuming services, but they should intervene in the decision-making processes.

These two conceptions were reflecting themselves in the different strategies envisaged to shape Iperbole, and their interaction was going to modify the initial character of the initiative, as we have seen it in the previous sections. In the secretary for Innovation's vision, the initiative was born as a way for the public sector to get to cyberspace first, before the private sector could commodify the local scene of networked information. This meant that the Council was going to grant access to the Internet as a right. He argued:

> What is the public interest? Until three years ago nobody would have thought that giving free Internet connections could be a Council's function. People expected us to build roads, public lights, and the like. But now this line is emerging, although among a restricted group as yet.

But the necessary introduction of more services and synergies with the private sector, together with the need to make the initiative economically sustainable in the longer term, were going to introduce a looser vision of the role of the Council. From the aim of granting universal access, the Council's mission was going to be re-defined as that of 'facilitating' access (Comune di Bologna, 2000), and the free Internet deal was going to be terminated.

An entrepreneur-friendly Iperbole?

As an academic interested in new media and a local politician, the Council Culture secretary represented for the original members of the Iperbole team, the right person who could replace the former Council secretary for Innovation, who in 1997 was increasingly detaching himself from the initiative. The CINECA director explicitly stated his willingness to take him on board: 'The assessore for Culture has started being interested in the project. I do hope he falls in love with it as the secretary for Innovation did'.

Indeed the Council Culture secretary was interested in the civic network's potential, and willing to make a contribution. In fact he gladly agreed to be interviewed and provided a series of truly interesting comments on Iperbole, some of which have already been quoted.

As the Council's appointee responsible for culture within Bologna, he had of course his own agenda, and this appeared mainly centred on the development of a cultural industry of small enterprises within the city, as mentioned in the section on Iperbole and the local economy. Therefore, he favoured a different approach to the management of the civic network, as his main concern was the excessive closure and bureaucratic character that Iperbole had shown until then. Opening up Iperbole to all those groups, especially made of young people, who could use the network to create opportunities and liase with other similar cultural entrepreneurs around Europe and the World appeared as a chance not to be missed. the Culture secretary was deeply critical about the 'purity' of the initiative that, in his opinion, was a weakness rather than a strength, something that limited the power and complexity of the civic network, and kept those groups out of it:

> If Iperbole keeps being 'pure' by excluding any commercial or entrepreneurial operation, then it has defined its function. But I am sure that the most interesting things will happen outside Iperbole.

The need to 'open up' the civic network was seen by the Council Culture secretary as extremely important also within a more general picture: 'To have some results on the social side, the Council alone is not enough', and

> I believe that Iperbole should offer a series of virtual spaces in which a certain freedom of action and movement exists. Otherwise it is going to be just a machine for Council services. We should define ample and flexible rules, and wait to see what happens before going into normative details too early.

The Culture secretary however recognised the importance of increasing the Council services offered through the Internet, and he was not against the city manager's PDS, although his vision was aimed well beyond the concept of Iperbole as a tool for service distribution. The project manager, well aware of the problems that closure and limitations of use had generated, agreed that

> We should open up to the commercial sector in order to let this become an element for the survival of the civic network. We should also address the other problem of letting ideology and active politics in, to bring the network closer to the actual reality of the city.

The Culture secretary's pragmatism aimed to promote what was seen as an imperfect but viable and efficient form of pluralism in Iperbole. Rather than trying to reach what he saw as an improbable universal participation, he wanted to encourage those who could and wanted to get in and make things happen, by giving them more freedom. This also meant to him that the Council's resources would have been better spent on creating specific opportunities rather than granting access indiscriminately:

> I agree on the fact that we will have to re-think the offer of free connections. Maybe we could identify some categories to help that would still be entitled to free access, but in general people should pay a subscription. This would let us be free to do other things

and use the money to finance and help those who are willing to use the network to create new jobs and expertise (Council Culture secretary, 1997).

However, only some aspects of the Culture secretary's vision were going to impact on the project. If on the one hand some opening up to private firms was going to happen, this was done in a much more bureaucratic and 'regulated' way compared to his expectations. It can be argued that the overarching bureaucratic character of Iperbole was extremely hard to change, and the freedom of movement and action within the network envisaged by the 'assessore' was perhaps too much for an initiative driven exclusively by an Italian city council. In the next section we will see how these new ideas from the actors involved contributed to change the shape of the civic network in the following two years, and how the several problems of Iperbole have, or in some cases have not, been addressed.

Towards the digital services city? Weaknesses and strengths of the new Iperbole

We have seen how the search for a solution to some of Iperbole's weaknesses, together with the closer involvement of some new actors, were putting a special emphasis on three main issues. The first was the need to develop and offer more interactive services, including those involving economic transactions. The second was to loosen up the rules and limitations of the original Iperbole, and let the private sector in. The third was to make Iperbole economically sustainable in the medium-long term. Some interviewees, as the project manager, the Council Culture secretary, and the director of Omega, also envisaged that, as Iperbole would become more complex and useful, and more resources would be available, an indirect consequence of these changes would be an increased participation from the public. This unfortunately has not been the case at least up to the date these notes are being written. But much was going on in the months following the beginning of 1997, and it is worth examining where the efforts were going and how the civic network was changing.

In terms of access policies, it was soon decided that the totally free connection deal could not be sustained, as mentioned by some interviewees in the previous sections. However, the overall spirit of offering a very cheap service, more affordable that the one provided by the commercial sector remained, and from 13/1/98 onwards all users were asked for a once-for-all contribution of 70,000 liras, the equivalent at the time of about 26 pounds (Comune di Bologna, 1998a).

However, three years later, in January 2001, the connection fee was waived again, as Internet connectivity was becoming cheaper and more choice was available in the marketplace.

The number of individual users at the end 1999 was 15,600. This was possibly much lower than it could have been expected, having grown in two years by less than 2,000 people. But it cannot be argued that this was just a consequence of the new connection charges. Also, a lower number of membership than expected did not mean at that time that Bolognese Internet users were not growing. As was happening in the UK, telephone companies had started offering free connections to the Internet too, where only the cost of the call would be charged. Therefore it is plausible that many people in Bologna could be surfing the net and possibly using some of the civic network facilities from non-Iperbole accounts. This is suggested by the steep increase of the daily 'hits' – or visitors – to the Iperbole web site. It is notorious that the number of hits gives a very superficial indication of the actual usage of an Internet site, and some of these contacts inevitably must have come from outside Bologna. However, the fact that at the end of 1997 the daily hits totalled 12,800 (Comune di Bologna, 1998b) and that two years later these had grown to 96,000 (Comune di Bologna, 1999a) certainly means that usage of Iperbole had developed considerably.

As could be expected, the provision of services – many still in an experimental phase – was the main focus of the efforts and investments within Iperbole. At the end of 1999 it was possible to apply for Council nurseries via the system, having some taxes calculated and reviewed, accessing the so-called 'general protocol' and check the status of planning applications and other Council-related bureaucratic tasks. Electronic booking of cultural, tourist, and conference services were being scheduled to be implemented during year 2000. An experimentation on locally-based Internet shopping and Council payments was being held through the web, and 700 families had been given a TV set-top box to connect to these services through their televisions. Because transactions had to be secure and reliable, research and experimentation was also going on in implementing digital signatures, and about 500 citizens had been given one (Comune di Bologna, 1999a).

According to the project manager's intentions, the elderly were also going to be offered specially tailored services through the participation in a European project called 'Senior Online', partly funded by the EC. The good reputation of Iperbole was again beneficial, bringing in new opportunities for the funding of development.

Figure 5.7 The 'Senior Online' project

In general, Iperbole was somehow managing to open up to business-related activities and to the private sector, but as was mentioned before, it was doing it still from the perspective of a quite bureaucratic organisation. The 1998 plan to establish a 'Virtual Industrial District' envisaged the creation of a directory, organised by categories of business, of standard information pages about local enterprises, available publicly, and a series of communication services mediated by the Council between firms or between firms and general public (Comune di Bologna, 1998c). The Virtual Industrial District in its conception looked more like an intelligent database than as a marketplace and/or a creative and competitive environment. The whole idea seemed well far away from the open and lightly regulated environment that the Council Culture secretary had envisaged for the benefit of innovative entrepreneurs.

With the new focus on services and economic opportunities, the discourse about Iperbole was changing accordingly. It is interesting to notice that new official documents were using expressions like 'The web from information to services' or 'the virtual community as an efficient tool for global communication and business' (Guidi, 2000), that would have been unimaginable just three years before.

Beyond the emphasis on services, other efforts were being made to try and keep the 'inclusiveness' issue clearly on the agenda of the civic network. Apart from the project to provide services for the elderly, some progress was made on the previously weak front of computer literacy initiatives. The establishment of a centre dedicated to promoting computer and Internet literacy was – and still is at the moment of writing – on the agenda. However, courses started being held in 1999 geared to users outside the Council itself. Ten courses for a total of 150 representatives of non-profit organisations within Bologna's province were held in 1999, and a series of seminars and initiatives scheduled for the year 2000. These courses, though not totally free of charge, were subsidised by Iperbole, costing each participant the equivalent of 40 pounds for a 30-hours module. Sponsorships and synergies were also being explored with several organisations and firms in the city, from the local branch of IBM to local computer clubs (Comune di Bologna, 1999b and c and 2000b). In particular, an agreement with ATC, the public bus company, and a local bank, meant that by buying an annual public transport pass one would get a complimentary Internet course, an interesting initiative to promote sustainable travel, both 'real' and virtual.

However, despite some increased efforts towards the promotion of access and the inclusion of citizens, too little was still being done, and it appears that this, together with public participation to decision-making, was becoming the weakest, least resourced point of the agenda.

Public access terminals, for instance, were nominally slightly more that in 1997, but still far too few. A total of 14 access points – often endowed of just one computer – were available, and out of the 14, 8 were working only on a part-time basis or were limiting access to specific pages (Comune di Bologna, 2000c). Considering that the director of Omega was thinking that 30 would have been a fair amount of public points in 1997, it is clear that resources were still very scarce in this respect.

Similarly, although the issue of 'Electronic democracy' was on the project manager's agenda, and nine special discussion forums had been opened in 1999, observations carried out in May that same year revealed a picture of very limited participation. At least a couple of these forums looked practically deserted, while the 'traditional' discussion groups, as mentioned before, saw considerable activity only in areas such as 'jokes' or the second hand market. It seems evident that the hopes of an increased participation of users attracted by services were not being fulfilled, and that this was a strategically weak and maybe neglected area, that needed much more work from the Council. A much greater involvement of a complexity of actors, both within and outside the municipality, would have made possible outlining a more detailed and efficient strategy, which could be 'grounded' more within Bologna's communities, in the way the senior planner had envisaged in his interview.

Conclusions

Some major points can be made to summarise the outcomes of the Iperbole case study.

First of all, the usefulness of case study itself, as it allowed an in-depth analysis which was clearly not possible through the instrument of survey. In fact, the Iperbole initiative had a high score within the contents-based investigation. However it has to be noted that the relative shallowness of the instrument of survey itself meant that the presence of certain facilities and characteristics within the civic web site produced a high score, regardless of how successfully – or unsuccessfully – these had been implemented, and what policies laid behind them. In a way, this implies that innovative initiatives of this kind, often initially accompanied by a considerable amount of hype, may present wide gaps between intentions, claims and actual implementation.

On the one hand it cannot be said that the Iperbole initiative was unsuccessful. First of all it fulfilled quite well the political purposes that it had been supported for, and further improved the good public image of Bologna's administration. It also created the conditions for further involvement of Bologna's team, especially the Council and Omega Generation, in European R&D IT projects that could turn out to be very beneficial for the city. From the point of view of how it managed to support social access to the civic network and the wider Internet, it has to be said that the free accounts initiative was certainly unique, and it contributed to attracting some members of the general public towards the telematics tools. A 'critical mass' of local users was reached within a relatively short time of the starting, and was certainly a most noticeable achievement. Iperbole in its first phase was motivated by the vision of public administration having the aims and duty to provide universal access to telematics networks. This was done within a discourse emphasising the divide between the functions of the public sector and the possible threats coming from the direct involvement of the private, commercial sector. Therefore all partnerships were shaped in a hierarchical way: the Council was the ultimate owner of Iperbole, and no-one else. This created the conditions for developing the initiative towards equal non-profit access, which was one of the purposes, and it 'educated' a certain amount of households. At the same time, though, it left out of the civic network a considerable amount of potentially dynamic actors.

Leaving actors out of cyberspace, that in 'real' space would certainly have played an important role, was therefore a weakness of the initiative. Although this might have helped to preserve a certain 'purity' of Iperbole and to avoid risks of early commodification of information and services, it also implied somehow that the Council was supposed to be perfectly able on its own to represent the city on the net. But the Council was not 'the city', and failure to involve a wide pool of actors – economic actors, certain pressure and political groups, alternative culture youth groups etc – in the shaping and management of Iperbole was proving harmful to the potential of supporting public discourse and generating more interest. Also, more involvement of economic actors would have meant the ability to generate some money and improve the medium-long term sustainability of the initiative, together with more resources to work towards access and computer literacy within the city.

Some problems were evident from the very beginning about fragmented strategies and actions towards the promotion of public participation and the support of public discourse within the city. The establishment of public debate areas within Iperbole could have worked if coupled with initiatives in the field that would be carried out in parallel. In other words, public participation purposes were not addressed by the Council with a coherent and unitary strategy that could take advantage of the Iperbole resource. The civic network was also a not very negotiated initiative within the Council itself, as only one office was driving it, expecting all the rest of the Council to follow suit. It was inevitable for the Council 'machine' to be extremely slow to adapt to new modes of communication and distribution of information. Although this could be seen as physiological, and only time could overcome these kind of problems, the hype about 'citizens having dialogues with the mayor' was going to prove hasty to say the least, and bi-directionality remained a serious problem.

The involvement of new actors brought a shift towards a more open and service-oriented Iperbole, certainly improving long-term economic sustainability, and making the initiative more useful and alive. However, the lack of fully participant partners – other than the Council's project managers – in the driver's seat could still mean that the potential of the civic network in terms of local economic development and public participation was not fully exploited.

The paradigm shift from an universal access granting initiative, mainly focused on innovative tools for local democracy, to a service-oriented digital city that facilitates familiarisation with the Internet, has not had apparently many consequences on low levels of public cyber-participation. The hopes that more services would bring more people in could be right, but would these people participate and debate anyway? Normal usage of the wider Internet does not seem to suggest this, and it somehow shows that Iperbole's management had – or had been forced to have – a relatively technologically deterministic attitude. The need for specific policies and initiatives, within and beyond Iperbole, for supporting public participation remained and meant that telematics initiatives must be part of wider, detailed strategies for urban and community regeneration.

Chapter 6

When Bristol Went Digital

Introduction

Finding a representative and significant case study of the digital city phenomenon in the UK has never been an easy task. The survey conducted within this wider study pointed the fact that although the number of Internet/web civic sites was high in the UK, the vast majority of these initiatives were not very interesting for the purpose of this investigation. Most sites, despite carrying promising titles such as 'Virtual Sheffield' or 'Wolverhampton Community Internet Project', were actually just promotional brochures for the city, targeted at non-locals, or relatively poorly compiled directories of local organisations and telephone numbers. Interestingly enough, several major cities in the UK were 'represented' by many different sites competing with each other and all claiming somehow to be there to contribute to the life of their physical counterpart. Bristol was just one of these cases in 1996, with six different 'virtual cities' – including a Council site – all offering low standards of both information richness and interaction (Graham and Aurigi, 1997, p.39).

However, during 1996 a partnership between Hewlett Packard, the University of West of England and Bristol City Council had been working at the creation of a new community-oriented web site for the city of Bristol, which was going to be officially launched in early 1997. In particular, one of the main promoters of this initiative, a senior researcher from Hewlett Packard Laboratories, had been presenting the project on the new digital city at conferences on urban themes during the previous months. The concepts behind this initiative were very relevant to the topic of this study, as Digital City Bristol Interactive (DCBI) was being envisaged as a truly versatile, community-oriented, highly interactive and participation-friendly project, supported by a really interesting partnership including a major global technology company like HP.

This chapter deals with the field research carried out on Digital City Bristol during 1997, and successive observations of the site. As with the previous case study, this investigation involved visiting Bristol and interviewing the major actors involved in the shaping of the initiative, as well as collecting available documents on Digital City Bristol. However, as is described later in the chapter, the project was at an earlier stage of development compared to Bologna, and only a few features had been implemented in respect to the original plan. Therefore, less information was available about it, including the fact the interviewees themselves had less to comment on.

The chapter deals with the initial phases of the project, defining the various actors involved in shaping the initiative and their interpretations of the project itself and of civic Internetworking. Then the narrative moves on to deal with the more specific topics of the relationships of the digital city and the real city in terms of support of public discourse, economic regeneration, and enrolment of users and the management of their access to information. The last section examines the evolution of the initiative towards a rather different model and conception with respect to its very early days, and the reasons that contributed to generate this shift.

Bristol

Bristol is the largest city in the South West of England. Once a famous port and centre for international trade, the city's economy became primary location for the aircraft industry and home of Rolls Royce's aero engines factory. The city had been trying very actively to attract fresh investment, and the presence of the technology giant Hewlett Packard – especially with the 'softer' presence of its research laboratories rather than production chains of hardware – was certainly seen by some of the interviewees of this case study, as a major asset.

On the local politics and administration side, it must be noted that Bristol City Council had become a Unitary Authority on 1st of April 1996, after the abolition of the former County of Avon. This meant that local government had to re-organise, and this brought problems to solve as well as opportunities to do things differently, embarking on a series of initiatives to improve the efficiency of the Council, in which IT had a central role (Bristol City Council, 1997a, p.1).

Bristol is also a well-endowed city as far as the presence of popular universities – and their students – is concerned. Two universities, the University of Bristol and the University of West of England (UWE) guarantee the presence of a good number of young people in the city, and there is a high demand for leisure, cultural, and nightlife activities, and in general a vibrant environment. UWE, the former polytechnic, is also one of the most research active and dynamic 'new' universities in Britain, and this was certainly a factor – as we will see in the next few paragraphs – facilitating the conception and implementation of innovative initiatives in the city.

Actors and visions for Digital Bristol: HP Labs start researching

These conditions were certainly important factors influencing the birth of Digital City Bristol. They established an arena in which IT industry, academia, and local government were all represented and willing to have a project in common.

However, an initial set of ideas and a vision for a specific project such as Digital City Bristol was obviously going to come from a specific promoter, or technological entrepreneur. In this case, it was acknowledged by all interviewees that the original idea for the design and establishment of a 'digital city' in Bristol that could be more sophisticated than the existing competing 'Internet directories' came from Hewlett Packard Laboratories, and in particular from one of its senior researchers. As the project manager and research coordinator from UWE put it, 'all of these projects have some idealist or ideologist promoting them, and we had [the HP senior researcher]'.

The role of the senior researcher, especially in the first phase of conception and life of the initiative was indeed central. A psychologist engaged in studying behavioural aspects of people using high technologies, he had been in talks about similar ideas with a colleague, from San Francisco Digital Media Center. This contact was created as HP Labs – whose central offices are in Palo Alto – had started liasing with the Digital Media Center to gain a better understanding of issues concerning digital imagery on the Internet (HP senior researcher, 1997). In the late 1990s, as the Internet seemed to be booming, as well as all sorts of business opportunities related to it, many IT giants that had so far engaged in producing hardware or software packages, were trying to adjust their strategies towards the production and management of Internet content. Microsoft Corporation, for instance, launched in that period its Microsoft Network portal, pages, and services, better known as MSN. Part of the industry was also trying to push ahead the new paradigm of software that could be loaded and used on demand, through the Internet, possibly charged on a pay-per-use basis, rather than residing on people's hard-disks. The conception and development of the innovative Java programming language by Sun Microsystems could be obviously seen as part of this strategy, as Java allows programmers to create applications that can be distributed and used over the Internet, on a wide variety of different platforms and operating systems. In such a climate for the IT industry and market, it was quite natural for a company like HP to be proactive in considering how Internet content could be effectively delivered, and through which applications.

Hewlett Packard could therefore see itself as a potential 'shaper' of innovative, content-rich websites as it recognised the need for a wide-ranging research approach in the field, in order to understand both the technical and behavioural aspects of Internet content delivery. Another reason why HP would end up considering the Bristol initiative as feasible, was its 'good citizen' policy, that is the willingness to give some benefits back to the community that hosts HP's offices (HP senior researcher, 1997).

Following the contact between the senior researcher and the Digital Media Center, the first intention had been to work on some kind of digital 'twinning' between Bristol and San Francisco, through the creation of websites. However, other factors played a major role within this early phase of the conception of Digital Bristol. One factor was the fact that the HP senior researcher, probably facilitated by his Dutch origins, had in that same period got acquainted with the digital city initiative that had established itself as a European paradigm for civic Internet: De Digitale Stad (DDS) in Amsterdam. He was fascinated and interested in Amsterdam's digital city success, and managed to invite the 'mayor' of DDS, Marlene Stikker, who gave a talk at HP Labs in Bristol. By this he convinced the top managers of the Laboratories that such an initiative would be 'a good investment' (HP senior researcher, 1997).

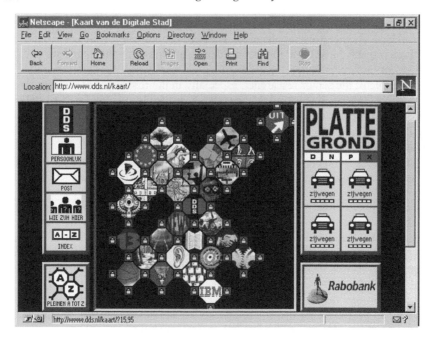

Figure 6.1 The 'urban' interface of De Digitale Stad (DDS) Amsterdam

It seemed clear to the senior researcher that the model to base the digital counterpart of Bristol on was DDS, as it had a strong emphasis on interaction and participation, and this was being recognised as an important component of the research:

> We saw the digital city as a possible research tool. The social aspects were interesting to us, those new ways for people to communicate. Our first thought was about facilitating communication in the digital city, rather than putting emphasis on information (HP senior researcher, 1997).

Reinforcing the role of DDS as the main inspiration source for what was going to happen in Bristol, HP Labs funded two bursaries for a Masters degree at Delft University of Technology, Faculty of Industrial Design Engineering, in the Netherlands. These were meant to support work directly related to digital cities and to layout the basic concepts, ideas, and the interface itself for Digital Bristol. The results of the study and the design by Sierd Westerfield were summarised in a project report for HP Labs titled 'Digital City Bristol Interactive', a title that also became the initial name of the initiative (DCBI). Coherently with HP's main interests, this piece of work was based on a case study on DDS, and aimed at envisaging optimal user interface and communication tools for Digital Bristol (Westerfield, 1997). Topics like the type or quality of information and contents, or implementation of policies to grant universal access to the citizens were not – or only very marginally – touched upon.

Figure 6.2 The early interface of Digital City Bristol Interactive (DCBI)

So it could be argued that the digital city in Bristol had been initially conceived as a minor research and development project initiated by a big multi-national company, whose main purpose was to study ways of creating highly interactive, grounded Internet sites. As Westerfield's report remarks: 'The fundamental question addressed in the Digital City Bristol research is: What are the opportunities for HP in the Emerging Information Infrastructure (EII) consumer market?' (Westerfield, 1997, p.11). Technical issues of software implementation, as well as on human-computer interfaces, were very much at the centre of the project. Another student from Delft had also been involved to study the use of Java programming technology within the initiative (HP senior researcher, 1997).

However, the person in HP Labs who had started DCBI, the senior researcher, had a wider vision for it, being strongly interested in the social aspects and implications of the establishment of an interactive digital city. A short paper published at the time on the HP Labs WWW site, in which different types of early digital cities are compared, kept remarking on the importance of people's presence: 'What a digital city provides is people' (Geelhoed, 1996) and of their interaction and empowerment in an electronic equivalent of public space, as opposed to an electronic version of shopping malls. The paper quoted Geert Lovink, a Dutch philosopher and critic, endorsing his claims that:

A lot of people in government and in business think that if they are just offering information, it is enough. But this is not the case. It is very important that the people can find an identity themselves with the media and that they are part of the computer network. That they are not just treated as consumers who are just buying something (Lovink, 1995; quoted in Geelhoed, 1996).

This interest for letting the embryonic digital Bristol actually involve and empower real people, and go beyond the limited scope of a human-computer interface research project, was motivating the senior researcher to act as a central character and facilitator, trying to generate support within HP as well as enlarging the circle of the actors involved:

> The first year of activity was about creating a network of people who could be interested in the idea, the idea being that it should not have been just a commercial place for electronic trade, but a meeting place.

So, to a certain extent, the senior researcher and HP Labs could be seen as two distinct actors in the Digital Bristol arena, insofar as the former had been successful in enrolling HP – his own employer – in the initiative by highlighting those aspects that could interest the multi-national company's R&D division. The HP senior researcher had 'sold' the idea to his employer by presenting an appealing interpretation of it, a set of potential benefits within something that could potentially become a much wider, all-round, and participated project. He has also made it economically very low-cost and viable, as this was seen within HP Labs as a minor project, and the costs were reduced to comprise a part of the senior researcher's time, the two studentships with Delft, and a small amount of equipment.

However, other important actors were to be involved in the management and shaping of Digital City Bristol, and their participation meant that different interpretations of the role and nature of the initiative were to be made, and these of course were going to have a deep influence over the actual way the digital city was going to develop. As we noted before, the HP senior researcher was looking for people to populate the forthcoming cybertown, and as the UWE research coordinator noted: '[the researcher] claimed that he was going to have the digital city designed, but to bring it to life he needed people from the city and from the university'.

The University of West of England

Of the two universities in Bristol, it was the University of West of England (UWE) which became involved in the project. A couple of reasons seem to explain this: physical proximity – which in itself is a fascinating issue to have to consider when it comes to speaking of projects involving cyberspace – and the dynamism of the management. UWE's campus is located out of town, within walking distance from the HP buildings, which facilitated frequent contacts between researchers and managers, independently from this specific project.

Above all, however, it has to be noted that UWE is a second-generation university, one of the ex-polytechnics that had found themselves 'upgraded' in status in the 1980s, facing the new challenge of re-defining themselves from just teaching establishments to being research active institutions as well. Some of these universities were working very hard and proactively towards increasing their research output, and UWE was certainly considered to be a very positive example of this. As a consequence, UWE was bound to be very keen on innovative research or R&D projects that could boost its image, prospective funding, as well as its internal research culture. It was indeed an ideal potential partner for Digital City Bristol Interactive.

HP had been interested in collaborating with UWE for some time. Years before DCBI was conceived, there had been 'a "famous" dinner between the Head of HP Labs and UWE's Vice Chancellor' where 'HP challenged the university' (UWE research coordinator, 1997). The HP senior researcher himself had been in touch with UWE before the digital city was conceived. HP and the university had also been in talks to organise some type of a 'media' event in Bristol that would 'bring potentialities together' (UWE research coordinator, 1997) in the city, generating new partnerships and new projects. The research coordinator was an important player in this respect, as she had been appointed by UWE as a manager who was expected to create momentum for research projects and consultancies. After these first contacts, including the already mentioned 'twinning' intention with San Francisco, the senior researcher started working on the new digital city idea, and went back to UWE to present these initial concepts to them, while the students from the University of Delft had already started working.

UWE got on board quickly, as the partnership with HP Labs was interesting and prestigious to them. The university's research coordinator defined the initiative as a 'very good research about the city of Bristol itself, that would enable us to give design and organisational recommendations about this type of initiative', possibly envisaging the university's ability to develop a know-how in innovative Internet sites, that could be exploited on the consultancy side of academic business.

Bristol City Council steps in

The other partner in DCBI had to be the 'city' itself, or rather the City Council. The UWE research coordinator took the initiative and contacted Bristol City Council shortly after the new head of IT services had been appointed there.

As already noted in the introductory section of this chapter, Bristol had recently undergone some major changes as far as local administration was concerned, becoming a Unitary Authority in Spring of 1996. This had lead to a process of re-organising and re-thinking about the ways the local authority should operate. Within this process of re-organisation, the idea of a specific initiative, or an 'umbrella' of initiatives involving information technology has started emerging, and was eventually named as 'IT in the Community'. According to Bristol's documentation for bidding to get a '1997 Local Government IT Excellence Award' this was

> an attempt to think creatively and laterally about how the Council could lever its existing and planned investment in information technology to contribute to a wide ranging programme of economic and community regeneration in the city. Key themes

included wealth creation, reducing unemployment, improving infrastructure, improving quality of life and empowering local communities (Bristol City Council, 1997a, p.1).

The head of Council IT services saw the proposal about the digital city as something that would fit well within this overall IT-based strategic vision for Bristol, and found good support from

> some of the leading politicians, who felt that there was a danger that the Information Age was going to pass a lot of people by in the city, because of economic barriers. So one of the ideas was to share information electronically and make it accessible (head of Council IT services, 1997).

The interesting thing to notice about the Council's involvement in DCBI is that the digital city was going to be seen as just a complement to an existing and wider overall strategic plan for IT in the city. So, if on the one hand Bristol City Council was interested in participating in the initiative, this was being considered as a piece of a bigger jigsaw, and possibly not the central one. This was clear from the fact that – being just a partner in DCBI and not even the most influential – the Council would keep developing and placing most of its information and service resources within its own web site. Although this should not be seen as a necessarily negative aspect of the initial development of DCBI, as it was fairly normal that partners would retain their own separate identities, it suggests that in this Bristol case the fragmentation of agency could be mirrored in the Internet, rather than be moderated by it. And that this could easily lead to a wide fragmentation of interpretations of what the digital city was going to be there for. However, as long as a focus on common benefits and opportunities could be kept, the initiative could start and progress, as we will see in the next section.

Why Digital Bristol? Interpretative flexibility in accepting and shaping the cybercity

The introduction of the several main partners involved in the digital city has already suggested how the Bristol case could be considered an arena in which several sets of interests and objectives were represented, and played a role. What opportunities could Digital City Bristol Interactive offer to a local authority, a multi-national high-tech company, and an emerging university? Why and how could actors who were so diverse converge towards the initiative? Also, to what extent were they really converging, therefore making the digital city viable and supported in the long term? The answer to the latter question will be better addressed later in the chapter, when the evolution of the project towards a certain degree of stabilisation will be dealt with. However, now it is time to have a better look at how the different meanings the digital city was embedding for the main actors involved could somehow combine to facilitate the birth and initial growth of the Bristol cybertown. A main theme and driving factor can be identified here: the emancipation and 'de-provincialisation' of Bristol – including the projection of a brand new image of the city as a forward-looking, dynamic and efficient place.

Networking to re-brand the city and its actors

As mentioned before, Bristol in the late 1990s was a city trying to re-define itself. We have seen how the local authority had assumed a new role, and the UWE too was working hard to re-launch itself as a proactive, research-based institution. Therefore a project like the digital city could be seen as a great opportunity for helping to establish a strong and forward-looking image for the actors involved. Certainly it was also a good chance to network for some of these institutions, and experiment with closer partnerships. The UWE research coordinator was very explicit about this by saying that

> this digital city is about getting the city's influential organisations together. There is a kind of new optimism in Bristol now, and what we are doing with the digital city is about right, using it to get key individuals and sections of the community to work together.

It is interesting to notice the clear interpretation from the research coordinator of the specific digital city initiative as something to 'use', a good catalyst for collaboration among the 'key individuals' in Bristol. She clearly identified networking, together with the more obvious research, as the two main UWE's priorities. In particular she saw the digital city as a great opportunity to get a deeper relationship with HP, as the visually-intensive character of DCBI was going to be well suited to the interests of an 'ex-polytechnic with a strong faculty of arts, media and design, as well as a strong computing section' (UWE research coordinator, 1997).

Although from a quite different point of view, the City Council also had a clear re-branding agenda for the city, the council itself, and its activities. Quickly embedding the digital city into the wider set of projects and policies of 'IT in the Community', and defining DCBI as 'a milestone', the Council was aiming to demonstrate 'very explicitly the potential of the new Unitary Council to make a real impact in the city through the innovative application of IT' (Bristol City Council, 1997a, p.2). Not surprisingly, the theme of efficiency was central among the reasons for the local authority to get involved, as could probably be observed in many other cases. The head of Council IT services confirmed this by describing the cybercity as a 'new communication medium that allows the Council to do things in different ways and do new things. And above all do them more efficiently and spend less money to run certain types of services'.

The 'vision' from the HP senior researcher of a participative, communication-intensive digital city close to the model of its Amsterdam predecessor, can be seen as yet another interpretation converging towards the general expectation of a less provincial, more lively and proactive Bristol. It has to be said that the senior researcher's vision would go well beyond a simple place-marketing ethos, or the consideration of the digital city as a good excuse for institutional networking. However, the will to provide a hi-tech space for facilitating collaboration and giving people and groups a certain degree of visibility could be easily seen – at least initially – by the other actors as a very coherent and desirable course of action towards the common goal of 'making things happen' in Bristol.

Generally speaking, the perception formed during the face-to-face interviews in Bristol was that all major actors had recognised the need to 'de-provincialise' the city and facilitate the emergence of a new image of forward-looking place. Image-making was paramount and was the real cohesive factor for the partnership, as the Council's 'IT in the Community' document stated openly:

> A number of the specific areas of work completed have done much to transform the image of the City of Bristol towards a leading edge, high technology city. The profile established for the City Council through working with credible partners such as Hewlett Packard and the University of West of England on Digital City Bristol has been very positive (Bristol City Council, 1997a, p.6).

Although the perceptions and interpretations of what DCBI could represent were very different among the several actors, a common perception of 'inadequacy', or rather of a city that needed to regenerate its own mentality first, was clearly present. The UWE research coordinator, for instance, when asked about the issue of commercial Internet Vs universal access, dismissed this as something not to be considered in the short-term because it was somehow too advanced for Bristol. She claimed that Bristol was behind other places and cities, that not much had been going on there – she would repeatedly use the expression 'to make things happen' as an absolute must – and that it needed to catch up to be competitive:

> It is too early to think about commercial uses, or Internet access issues. If we had these things we would not know how to use them. People would talk to people in Newcastle, because the tradition there is better.

However, as we will see later in the chapter, the common goal of 'revitalising' Bristol would end up being effective in putting together different interpretations only for a limited time. It provided a good reason to start the project at all, and to work in partnership, but it would not be able to address the main divergence of visions for the digital city for the long term. As we are going to see in the next few sections, beyond the re-branding of Bristol, the expectations about specific aspects of how the digital city should work could be simply opposite, within the managing group. This would eventually generate the need for a radical re-definition of the initiative.

Public space with no public place: a fragmented digital city

Was Digital City Bristol an initiative aimed at creating a public 'cyberspace' in the city, a site where public discourse was going to be supported and have crucial importance? As we have mentioned before, the preliminary studies and plans made by the HP senior researcher and Westerfield seemed to claim that DCBI was a 'people to people' project. Another HP paper had defined this aspect as the focus of the whole R&D project (Geelhoed, 1996b), and Westerfield had remarked that

> The main theme is people to people communication (…) The DCBI will be the main site for Bristol. It will be the place where people can get information on the city, and a place where they can communicate with other people in the city (Westerfield, 1997, p.12).

HP had indeed based all its presentations about DCBI on the explicit link with the project of DDS Amsterdam. In a short study published as an internal HP Labs document and on the World Wide Web, different types of early digital city had been compared concluding that the most interesting was indeed the Amsterdam case, and arguing that:

> DDS has also been a major source of inspiration for our thinking. We judged three aspects particularly attractive: novel ways of representing the information space, enhanced navigational clues and last but not least the social aspects of the city. Residents of the digital city can communicate with each other via the digital city computer interface as individuals or as groups, thereby energising the city (Geelhoed, 1996a).

Clearly this vision was the course of action envisaged by the internal HP project report, and indeed by the initial design of Digital Bristol's interface and functionality. The digital city had been designed by HP Labs to look like a harbour, with an explicit reference to Bristol's past. The harbour was organised in 'piers' with virtual 'boats' moored to them. Piers and boats were there to host a wealth of different WWW sites that could be owned and run by community organisations as well as private individuals. The piers would be thematic to present a coherent and easily searchable organisation of the many informational 'fragments' present. The several functions available through this interface were activated by Java routines, which would have supposedly allowed a very high degree of interactivity, such as real-time zooming and moving within the 2D virtual environment. Communication functions were planned as a crucial aspect of the digital city. The HP Labs project report claimed that: 'DCBI will provide newsgroups as the basis for a discussion platform. The subjects for the newsgroups have not yet been decided' (Westerfield, 1997, p.36). But one the main challenges that HP Labs – but above all its senior researcher – were willing to face was working on the creation and implementation of a more intuitive and easy-to-use electronic environment to allow exchange of ideas among citizens. A whole chapter of the HP report was dedicated to the conception of what had been called the 'multimedia notice board' for the digital city (Westerfield, 1997, pp.36, 45), a graphics and multimedia-enhanced area for the members of the community to discuss, post and exchange files, announce events etc.

However, neither the innovative notice board, nor the newsgroups would end up being available at all in Bristol. The 'vision gap' between the HP senior researcher's open, interactive, lively, and communicative conception of the digital city, and the way the other partners were expecting DCBI to be, was extremely wide. In general, there was a strong concern for the problems that allowing free speech in the digital city could generate, and the majority of the actors involved shared this concern.

The City Council was declaring an interest in encouraging the development of an inclusive information society in Bristol, as we will see in the section on end-users development strategies, but mainly from the point of view of allowing the majority of citizens to use the new services and information. Access to the technology was indeed seen as an issue and it was going to be addressed. Communication-wise, however, the municipality was only concentrating – and not even too strongly – on providing councillors with email addresses, as the head of IT services noted:

> We have fourteen or fifteen councillors who are really enthusiastic. They have PCs at home connected to the council, and they are using email. Now it is a pilot scheme, but I hope next year we will be able to extend this to all sixty-eight councillors. It is a step towards electronic democracy. This is the kind of vision we have.

Indeed, the Council's strategic document on 'IT in the Community' despite outlining a series of initiatives and prospective benefits on information provision, education, and economic development, had basically no significant mention of any governance-enhancing effects of the information society in Bristol, and no plans for increased public participation or consultation through the use of IT.

Also, however genuine the head of IT services's vision for some kind of e-government trajectory for Bristol could be, thanks to increased use of email, the acceptance of new ways of working and dealing with the public was generally pretty low within the council. Those 'fourteen or fifteen enthusiastic councillors' seemed to be an unrepresentative minority to other interviewees. For instance, an economic development officer from the Planning department said:

> Different people use the system in different ways. Everyone in the city council has an email address, but many people still refuse to use it. Many do not reply to emails. Some even say that they do not like technology and are not willing to use it.

Another important thing to notice is that Bristol City Council's policy, as far as Digital Bristol was concerned, had never been that of migrating council services or facilities – such as the councillors' email – into the digital city. Bristol City Council had its own well-established WWW site, which was linked – and presented in a prominent position – by DCBI, but nevertheless was completely independent from it. So, the limited e-communication plans of the Council could well be carried forward outside the digital city.

On a similar 'line' was the attitude of the local press about the launch of DCBI. The local newspaper *The Evening Post* had been publicising the event with a positive review, but had been exclusively concentrating on the information 'broadcasting' benefits of the new medium, ignoring completely the bi-directional communication potential that the digital city could have embedded (Onions, 1997, p.11).

So, despite the initial drive from the HP senior researcher towards a lively and participative digital city that could have been seen as an evolution of Amsterdam's

DDS, the partnership behind DCBI was really not that keen on making it a cyber-Agora. Plans had been made to try and embed some debate facilities, but already with the clear awareness in mind that DCBI was not meant to be a place for free speech:

> With HP we decided in the pilot phase that we would not do anything which is completely free speech, as HP did not want to be associated with something that could cause an embarrassment. However, we are experimenting with some kind of newsgroup in a controlled and moderated form (UWE research coordinator, 1997).

The UWE researcher who had been assigned the webmaster, or in this case 'harbourmaster' role within the project, dealing with the day-to-day business of developing the digital city, confirmed that

> we are having someone developing some notice boards for DCBI, but the management committee do not want to have them as a real-time uncontrolled thing. So, each notice will have to be checked before it goes online.

Above all, evidence seems to suggest that the free-speech sensitivity problem was just an obstacle reinforcing, rather than generating, the trend towards the shaping of a different and less communicative digital city. In fact, if bi-directional, people to people communication was seen as a priority by the HP senior researcher, it was clear that he was isolated with his vision. The UWE research coordinator, when asked about public participation and public discourse issues, replied very uncompromisingly that 'the digital city is not about that. Mostly, it is about other things'. She would also mention explicitly the Council as having a information broadcasting target:

> The City Council is backing this specific initiative, among the several Bristol 'indexes' because of the involvement of local community groups and health care. Because it is about public information provision, it meets their public agenda. It is very different from the Amsterdam case.

So, shaping the digital city was a task embedding conflict as well as convergence and partnership. Above all it seemed to embed a good degree of 'fragmentation' of visions, when it came to deal with developing specific aspects of a low-funded, resource-thirsty project. The result of this was that the notice board that the webmaster was talking about was never fully implemented, and that despite the claims from the HP senior researcher and the UWE research coordinator herself that newsgroups would have been launched soon, none of these facilities was yet present in 1999, two years later, as noted by Annelies de Bruine:

> At the moment the web site is a good information source for Bristol. It is not very dynamic; more interaction can energize the digital city and make users come back to communicate with others. The chat/forum on the music pier and in Erik's Café is a modest start, but more newsgroups are needed to make the web site a dynamic and more interesting place on the web (De Bruine, 2000, p.122).

It cannot be argued, however, that DCBI was denying completely all forms of 'grassroots' expression. Indeed, the 'piers' and 'boats' were there to get populated by community groups and individuals who were given space for self-representation on the World Wide Web, and under the Digital Bristol umbrella. So, in a way the digital city was being shaped to become a highly diverse place. The problem, as far as encouraging public discourse was concerned, was that all the users – we could say the e-citizens – of the cybertown were able to become broadcasters and advertisers, but not to engage in debate within a public cyber-space. People were 'talking' to the world, but not necessarily together. Digital Bristol looked increasingly like a city made of unconnected – or connected only by the graphics of the interface – individual souls and fragments. And these fragments of course included the main partners working on the project: the Council, the university and HP.

The actor who was still proactively trying to establish some communication within DCBI, the HP senior researcher, ended up, as a consequence, acting as a 'fragment', and taking individual initiatives to make space for the exchange of ideas and interactivity. He was tolerated but not necessarily actively supported by the rest of the management committee. This meant that a new 'building' appeared in the virtual harbour, called 'Erik's Café'. This seemed to be strongly inspired by DDS, even graphically. It was meant to be an interactive, multimedia-intensive space for meeting. In its extremely short life it had been equipped with a MP3 jukebox for digital music, a reading room, and video clips of Bristol-based events. Also, in 1999, following a conference on 'Music on the web' held at HP Labs in February, a new 'music' pier was created and a chat/forum was implemented within it, targeted of course at music enthusiasts. Annelies de Bruine, another Dutch person recruited by the senior researcher, was the officer working on both the Café and the Music Pier, but HP Labs' commitment towards this was clearly very limited, as de Bruine was employed on a short-term contract and was soon going to move on to other occupations.

These sporadic attempts at actually establishing some form of public space within the digital city remained isolated within the wider context, and had a short life. Erik's Café, despite formally occupying a space on the main pier for quite a long time, had rarely been active and accessible, and disappeared completely with the third re-design of the interface.

Interactive? Better think twice...

The interface of Digital Bristol, as mentioned already, had been initially conceived to exploit a 'harbour' metaphor, and above all to make a relatively heavy use of Java routines that would allow end-users to navigate the information space effectively and still keep a view of the whole. The harbour metaphor, together with the Java-enabled navigation tools built into the digital city, allowed visitors to zoom in or out within the scene, and enjoy a sense of place. It was also considered to be an important feature that all inhabitants, individual users, community groups, and institutions, were sharing the same information space and 'graphically' lived together (Westerfield, 1997, p.28).

This approach focusing on what for the time was a high level of sophistication, requiring relatively high specification computers, was however not seen as appropriate by everybody in the management committee, and the interface was quickly re-designed in late 1997. The 'version 2' of the digital city kept the harbour metaphor and the relative imagery, only slightly changed, but dropped most of the Java contents to retreat to plain HTML (the basic World Wide Web language).

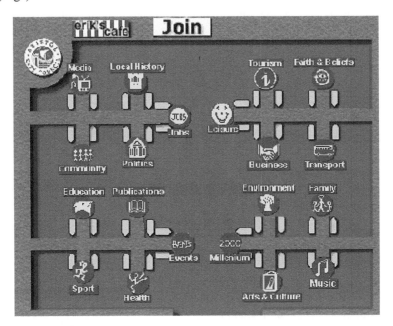

Figure 6.3 The Java-free version of Digital City Bristol

Again, the adjustment and re-designing of the interface was a clear sign that the DDS paradigm was not really acceptable to the other partners. The head of Council IT services argued that the Council was receiving complaints from most of the users about the ineffectiveness and slowness of the Java site, and the UWE research coordinator described this situation highlighting the conflict in vision and aims that had been developing:

> There are these two different layers of technical research and 'making it happen'. People who are interested in the technical aspect do not want Java to be removed, while those who are trying to use the digital city get frustrated.

This decision to get rid of the Java-based interface by the management committee meant that even the name of the initiative itself would go through a re-definition. What had started as 'Digital City Bristol Interactive' or DCBI, became simply 'Digital City Bristol', clearly stating that the focus on interaction was not going to be crucial any longer:

> At first it was called Digital City Bristol Interactive, but we had to drop the 'I' because we are aware that it is not very interactive. The whole idea of 'Eric's Café' was to have a chat place, and again that has not happened (head of Council IT services, 1997).

Indeed, in the year 2000 a further and yet more radical redefinition of the digital city took place, with the addition of new actors in the management group. These major changes led them to abandon the 'spatial metaphor' interface, getting rid of the 'harbour' concept, and adopting a far more traditional newspaper-like façade.

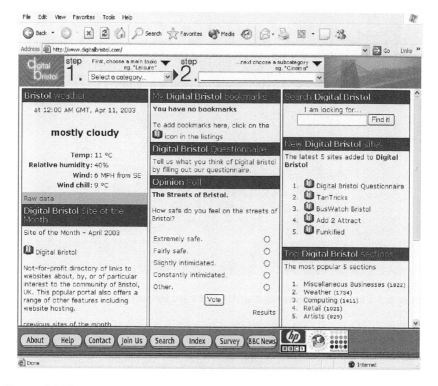

Figure 6.4 The new, non-urban interface of Digital Bristol

This in a way was the sign that the initiative had completed a cycle to re-define itself as an information broadcasting facility about Bristol, dropping the emphasis on electronic public space altogether. Digital City Bristol yet lost another word in its name. 'City' was dropped to leave it as 'Digital Bristol', something very different – if not opposite – from the DDS-like environment initially envisaged by the HP senior researcher.

Integrating physical and virtual strategies? DCB and the planning and economic development of Bristol

What we have seen so far could already suggest that the initiative was not meant to have a bearing on the planning of Bristol at all levels. The non-central role of the City Council in the management and development of Digital Bristol, as well as the lack of interest in going beyond the shared goal of shaping up a digital city for networking and self-promotional purposes, certainly played a significant role in keeping public participation – as we have seen – and planning issues in general out of the agenda. An interesting 'between the lines' indication of this can be drawn from the research coordinator's considerations on why and how UWE could find DCB relevant:

> HP envisaged that the next generation of computers would rely on people being visually aware, and not just numerically. UWE is an ex-polytechnic with a strong faculty of arts, media and design, and a strong computing section as well. This possibility of making the two communities work together was seen as very desirable.

What is interesting in this sentence, apart from what had been said, is what had not been said, or recognised. Indeed, UWE had – and still has – a proactive faculty of the Built Environment, with a strong focus on planning. However, no mention of any relevance of DCB for UWE's planning section came from the research coordinator's analysis of the situation. Indeed, no connection or specific interest from the planning researchers and lecturers of UWE towards the digital city could be observed. In other words, the digital city was a computing and media exercise.

This impression was confirmed by the visit and interviews in the Council, that revealed how weak – if it existed at all – was the link between planners and IT strategies – and people – in the city, and how little the ICT initiatives were being considered as relevant by those who were supposed to deal with the economic development of Bristol. It has to be said that the Council was supposed to have a vision linking its ICT initiatives to planning and, in particular, economic development strategies in the city. The head of Council IT services, in another document produced to present Bristol's strategies for IT in the community, had written that

> the City Council 'IT in the Community' project is an integral part of wider City Council strategies and initiatives, particularly those related to economic regeneration, community involvement and the development of productive partnerships with other sectors (Bristol City Council, 1997b).

However, if the IT section of the Council seemed convinced that the municipality's IT strategy, and indeed the digital city that was always mentioned as an important part of it, was going to boost Bristol's economy, no indication of this was coming from the economic development section within the same institution.

The Economic Development Strategy of Bristol City Council had been outlined by a document published in 1995, dealing with a 4-year long programme. This document presented a quite traditional approach to regeneration, or anyway a definite lack of consideration for the potential and impacts of new technologies. The 44-page long document scarcely mentioned ICT in relation to the city's development and management, apart from some brief and superficial remarks (less than a page) on the importance of the presence of a growing media and high technology industry in Bristol (Bristol City Council, 1995, p.16), however this was not associated with any concrete references to ICT implementation.

It has to be acknowledged that the economic development planner who was interviewed for this case study, was supportive of an ICT initiative that could affect Bristol's economy, but this, rather than Digital City Bristol, was a privately run project called 'SmartWest'. SmartWest was considering putting Bristol-based firms on the World Wide Web, and was linked to the so-called Internet Business Park, set up by HP and 'run by mail marketing as a completely private initiative' (planner, 1997). So, while the IT department of the City Council was hailing the ICT strategy and the digital city as a crucial part of innovation for Bristol's development, the economic development section of the same municipality was relying on a totally different, and completely private, initiative. SmartWest however was not a priority for the planner and his colleagues, and when asked more precise questions about it, he seemed to have a very limited knowledge of what the initiative was meant to achieve, and how. Moreover, the Council's economic development staff were at that very moment refurbishing their own WWW site to provide information to businesses. So, the digital city, and what it could have offered was basically ignored. The economic development planner was quite explicit about this lack of interest, saying that

> at the moment I could not honestly tell that we are promoting the information society to businesses. Not that we would not like to, but it is a matter of priorities and resources. I spend 5% of my time working on the information society.

It has also to be noted that the main focus for Bristol's economic development strategies was – as was the case of many other cities – on marketing the area and attracting external investment, rather than trying to generate some form of endogenous economic rebirth. The general vision about the World Wide Web and the Internet, and what it could do for business in Bristol, was very much focused on place promotion and provision of information for the outside world:

> From the point of view of us as users there is the way we promote information through the Internet and the web. It is making information more accessible to people who are overseas and can manage property searches, for example. There are all sorts of potentials (planner, 1997).

This perspective limited the vision and interpretation of ICT implementation in general as ways to broadcast information from Bristol to the rest of the globe, thus contrasting with the community-based character that the digital city was supposed to develop. In his interview the head of Council IT services claimed to feel this tension between the 'global' ambitions of the economic development strategy and the needs of local small enterprises. So, despite the fact that in Bristol there seemed to be no strong concerns for local businesses to be represented in the digital city

(HP senior researcher, 1997), a lack of a proper and shared strategy towards encouraging local firms to exploit Digital Bristol was in fact stopping this from happening. A manager from HUB45, a Bristol-based Internet company that had been invited to become part of the management team of DCB, sent some interesting comments to the UWE research coordinator in October 1997. He lamented exactly the lack of a more 'grounded' ethos of the digital city, beyond the promotion and networking exercise, stating that

> if ways can be found of making it easy and cheap for small businesses with rapidly changing offerings to put them before their local clientele much more quickly than by press or broadcast publishing; if sports, entertainment and leisure concerns can provide short-notice details to their customers; and if matters of public concern can be debated widely in real time then there will be a genuine enhancement to the potential quality of life of the region's citizens (HUB45 manager, 1997).

More generally, those responsible for the urban planning of Bristol were completely unaware of the potential and indeed the impact-related issues that the emergence of an urban virtual space could have generated. It can be argued that the general attitude was that of considering the issue irrelevant mainly because of a reactive, rather than proactive approach towards physical and functional planning:

> There is no immediate concern about this [the emergence of digital city initiatives]. Planners can identify the themes, but cannot see them happening yet. With a lack of hard evidence of what the implications really are, they have difficulties of putting this into planning policies (planner, 1997).

So, the chance of proactively using the digital city for innovative planning actions was completely overlooked, considering ICT as a phenomenon totally detached from planning, which could in the future have generated problems or issues to react to. The economic development planner identified the 'wiring' of the city and the emergence of electronic communication as something 'big' that would have a heavy impact on people's work patterns, future office space requirements and infrastructure planning. However, this perspective was indeed far too 'linear': new technologies happen, produce impacts, and planners have to react to these with policies. No consideration was given at all to the fact that planners could decide to step into the process of development of local ICT, and contribute to its shaping. Planners were keeping out, busy with other matters.

Populating Digital Bristol: constructing the virtual citizens

It has been mentioned in the section on communication and public discourse that although bi-directional exchange among citizens had not been supported strongly enough within the digital city's management team, the theme of making the information society inclusive and accessible had been somehow addressed. After all, one of the main characteristics of Digital City Bristol was its aim to 'host'

WWW sites created by community organisations as well as individuals living in the city and, to make this successful, access had to be encouraged.

In general of course it was everybody's concern to create a body of users of the initiative. On the surface, the target for the users' profile seemed clear and agreed within the management team. Digital City Bristol was aiming at mainly 'community groups', providing free space for individuals, but clearly with a minor emphasis with respect to what had happened in Amsterdam. The HP senior researcher saw the growing involvement of community organisations, that were getting free web space within the cybercity, as a winning point of DCB: 'There are a lot of community voluntary groups that you would not find easily in many digital cities. Two piers are already full with community groups'.

However, more deeply the different visions of the digital city, and therefore different interpretations on who the end-users should be and what they should be allowed to do, produced a 'fragmented' approach towards the 'construction' of the digital citizens and above all to the strategies to boost access to the virtual environment of DCB.

As seen in the literature review, the three main topics of universal/social access to ICT in the city, the problems of coping with IT literacy levels, and the barriers generated by age, gender, ethnicity and diversity in general form the framework employed to analyse this aspect of Digital Bristol's development and policies.

Access for all or an elite playground?

The theme of universal access to urban IT infrastructure appeared to be a strong commitment in the City Council's IT-related agenda, and indeed the municipality was the actor that had planned a course of action to allow disadvantaged citizens to access the Internet. The vision was that 'IT infrastructure is increasingly as important as roads and buildings' (Bristol City Council, 1997a, p.6) and therefore access to it could not be left only to the well-off. The head of Bristol's IT services reinforced this in his interview by saying:

> Why should people that are economically disadvantaged be furtherly disadvantaged? This is why people should be able to access Digital City Bristol. The social inclusion argument is really a big one. The fact that my children can use an expensive computer with multimedia encyclopaedias and some of their schoolmates cannot, does not seem right to me, and it does not seem right to the Council.

In practice, this aiming towards an inclusive local information society translated into one major policy and action from the Council: the funding and establishment of public access terminals, freely accessible computers located mainly in public libraries. Initially the installation of sixteen terminals had been planned, of which fourteen were going to be located in local libraries (Bristol City Council, 1997a, p.4).

Bristol City Council was serious about this, and in fact in the following years the number of public access points in the city grew considerably, to reach a much better penetration. In June 2001, four years later, the number of public computers or kiosks had risen to eighty-five, located in forty-seven different public or semi-public spaces in Bristol, twenty-seven were in libraries, ten in community centres, and ten in other buildings such as museums or charities (Digital Bristol, 2001a).

This was of course a very relevant move towards facilitating access to the digital city itself, but it has to be noted nevertheless that it was an isolated action towards boosting inclusiveness, and not much more was being done to address other aspects of the same problem by a wider strategy. Public access terminals can be important to grant access to the Internet and the local intranets, but because of the nature of the shared machines offered, with all the time and operative limitations, they can allow citizens to read materials and access information, but they can do little to allow people to self-represent and participate. This was even truer in the case of Bristol, where the self-representation in the digital city was supposed to be enabled by the creation and maintenance of individual or organisational web pages, something that could hardly be done from a public terminal. So, access was being encouraged by the Council, but mainly in the form of an ability to read information. Delivering content, or in other words 'broadcasting' to citizens was clearly going to be the main emphasis of the initiative:

> Developing local content for local people is an important part of the project and significant progress has already been made at several of the community-based Public Access Points. The City Council is particularly keen to work with other agencies to enable public access to a wide range of electronic content, and is also exploring opportunities of working with the Benefits Agency, the Employment Service and other government agencies (Digital Bristol, 2001a).

It could be argued that here lay one of the reasons why the 'residency' of individual web sites – in the harbour's boats – was never a success. The webmaster was pointing at 'English reserve' as the main cause of the lack of popularity of 'boat' spaces, and indeed he had a point in saying that 'People do not know what to put in their home pages. I have been doing my personal one for a while, and I have the same problem' (webmaster, 1997).

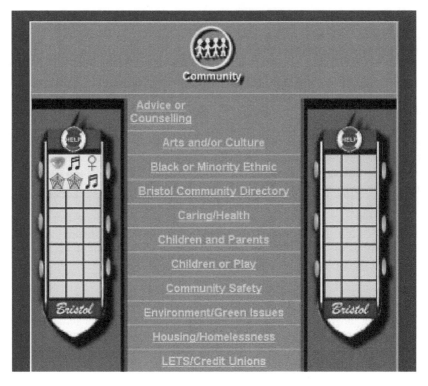

Figure 6.5 The empty 'boats' in Digital City Bristol's harbour

However, the difference between the relative success of individual self-representation in Amsterdam's DDS 'houses', and the deserted Bristol boats could also be explained by the difference in the emphasis on interaction and two-way communication in the Dutch case, and the growing dominant broadcasting ethos in its British 'equivalent'. Whilst organisations and charities could well be interested in the option of 'broadcasting' and opening their own virtual shop window, and indeed they would have been in general better resourced than individuals to take the opportunity, private citizens were offered a system that as a matter of fact recognised and placed them as audience, not participants. The webmaster seemed to acknowledge that a more comprehensive strategy for boosting access was desirable by saying: 'Maybe providing accounts to the community groups could be a key element to involve them, but at the moment they have to buy them from ISPs'. The HP senior researcher was also insisting on the bi-directional deficit of Digital Bristol as a possible key factor for the lack of involvement and the low taking-up of 'virtual residencies':

> Another thing that could make the digital city more attractive and encourage people to make their homepages is giving people ways of communicating with each other in the same way you can do it in Amsterdam (...) the advisory committee will have to decide whether to remove this aspect from the digital city or make it a reason of success.

However, as we have already seen, none of these facilities had been or were going to be implemented, and the setting up of the public access terminals looked like an isolated action, rather than a part of a wider strategy.

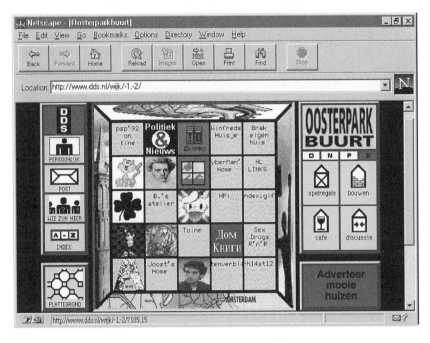

Figure 6.6 The virtual residencies in DDS Amsterdam

Some of the reasons for the significant yet limited action by the Council could surely be found in a relative weakness of British local authorities and in a consequent lack of money to invest. Bristol City Council was also too young to have been able to gain a relevant position within the Europe-wide circuits of cities, such as Eurocities or Telecities, and this was limiting strongly its ability to bid successfully for EC funds. When questioned about the chance of implementing wider policies and actions for boosting high-quality citizens' access to the Internet, such as giving residents free accounts to be used from home or subsidising the provision of computers for low-income households, the head of Council IT services was very explicit about the limits they had to work within:

> We are in a crisis of public funding. I do not really see how we could ever afford to offer free Internet to the citizens. We are having enough trouble funding nineteen or twenty public access points.

But lack of funds was not the only factor affecting the overall attitude of Digital City Bristol towards the encouragement of citizens' access. It is also important to remark that this specific digital city was not a council-driven initiative, and that the other partners could have done their bit towards encouraging more inclusive usage.

Again, radical differences of interpretation of the nature and role of the digital city were going to influence the shaping – and the non-shaping – of the strategy to attract end-users.

If Bristol City Council was in fact keen to encourage universal access, although to a somehow one-way digital city, UWE's agenda seemed definitely diverging from this. The strong interpretation from the university's research coordinator of the digital city as an innovative way – and pretext – for influential actors in Bristol to liase and work together, was in fact clearly implying that universal access was not an important issue that had to be dealt with by the owners of the initiative:

> Everybody would like to have free access, but the Council is very impoverished and does not have the resources to invest in similar things. The UK government has claimed to want to invest in schools. The possibility of getting connected through the electrical wiring and the electrical companies could radically change things. But the digital city is not about achieving this. It is more about getting organisations and individuals to work together.

Indeed, the research coordinator was seeing this project as mainly an environment created for an elite of executive people within the city, who really needed to communicate effectively and broadcast their decisions, whilst the majority of citizens could be informed by this or other more traditional – and one-way – media:

> I am not saying that this has to replace other media. It adds to them, but the radio stations could contribute to communication. It allows people who have responsibility to communicate with people who can disseminate messages.

So, it seems evident that the tension between the 'broadcasting' digital city and the more complex 'participative' one that was noted in the section on public cyberspace and participation, was really a central, crucial theme within the Bristol case, and that the first paradigm was winning over the latter for a series of reasons involving interpretation as well as financial constraints. Broadcasting of course implied the existence of broadcasters, who were going to be an elite of individuals and organisations empowered to use the digital city to its full potential, and convey messages to the wider audience of the citizens, for whom access was mainly the ability to read and retrieve information. A three-tier structure seemed to take shape, in which managers and key institutions had control over the cybercity, community organisations were empowered to have a voice, and were offered some training, but had no control over the shaping of the environment, and citizens ended up as audience. Consideration at the level of the community organisations can further enrich the understanding of this model.

What really is 'community'? Selecting who accesses help and training

Although the 'community' orientation of Digital City Bristol was recognised by all actors involved as one of its major assets, it is worth considering what 'community' was going to mean within this specific context. The word itself can be used in several different ways, so that it can refer to a 'holistic' social entity corresponding with the whole city – as it was used by the Council in their 'IT in the Community' manifesto, or it can refer to particular social 'fragments' of a strongly diverse city. It should be noted, also, that generically defined 'community groups' were supposed to get the benefits of free training, offered to them by UWE (HP senior researcher, 1997). So, what community, or communities, are we talking about in the case of Bristol?

Despite the fact that Bristol definitely is a very diverse city, hosting for instance a numerous Caribbean and Asian minority, as well as a town with two universities and a consequently large student population, the perspective on 'community groups' in DCB had initially a very clear emphasis, and it was relatively narrow. The webmaster helped define this. 'I say community groups broadly speaking, but they are mostly charities that involve arts and culture'. On the one hand, this is to be expected, as the project itself had to find leverage somewhere to kick off its offer of web spaces, and arts groups that would have been naturally interested in experimenting with new media could be a good choice. Needless to say, HP had a lot to learn from these multimedia experiments, and it was keen to encourage them. This interest was evident for instance in the 1999 conference on 'Music on the web', held at HP Labs and openly linked with the ongoing experience of Digital Bristol and its virtual spaces (de Bruine, 2000, p.121). Some institutions had also been approached, above all Avon Health Authority had been offered training to become an information provider within DCB.

Certainly a large number of groups were considered for help and some form of training. As the webmaster said in his interview there was a mail shot geared at 250 different organisations in the city. However, the fact was that everything was eventually dealt with by one individual, the webmaster himself, who had complete responsibility for organising and physically providing advice and training. This shows the limited nature of the concrete perspective on the involvement of communities. This actually reflected, again, the interpretation of DCB as a research project that both UWE and, to some extent HP, had. A one-man-band managing the involvement and literacy of people could hardly be enough for even a small research project. It was obviously too little for a growing digital city. Indirectly confirming this, the UWE research coordinator was very focused and precise in who UWE wanted to include in terms of communities:

> We do not want to involve everybody. This is still a pilot phase, and we just want to work with three different groups, and do it intensively. The first one is the 'health pier' and we have worked hard and long with the Avon Health Authority. This is a major success and evidence that digital city Bristol produces results (...) The second pier is the 'community groups'. We encourage them to think that they need a presence, and once they have a presence to monitor what actually happens – what they do, what it takes to get involved. The third group are businesses.

So, although DCB's main aim was allegedly targeted at communities, linking it to the wider Council IT strategy as well, the limited character that the initiative ended up having, and the difference in the interpretation of what it should be about,

narrowed the practical application of this vocation. Ethnic minorities were not in
the picture at all, and were never mentioned by any of the interviewees – or indeed
the documents obtained – and other sections of Bristol population, such as schools
and pupils, were identified just by one actor (HP senior researcher, 1997), but not
addressed for lack of resources and, possibly, interest. It can be argued that Digital
Bristol was opening up to an institutional or cultural elite that was instrumental to
its immediate success as a research operation. It had defined the restricted
environment it had to operate in its early stage of development, which could well
be seen as a sensible and viable approach, but at the same time it had not defined
any medium and long-term strategy towards involving the complex mosaic of
people inhabiting the city of Bristol. This was clearly denounced as a problem of
the project by Annelies de Bruine, who highlighted the lack of participation of
more citizens in publishing and indeed shaping the digital city:

> At the moment people see Digital City Bristol as a top-down organisation, although that
> is not the intention of the partners involved in the project. More citizens should get
> involved in building and maintaining the digital city. The management committee
> should think about a policy how to involve the people of Bristol more (De Bruine, 2000,
> p.122).

But Digital Bristol was going to further transform itself, as it has been already
mentioned in the section on interactivity, towards a more static and traditional
model of civic website.

Redefinition and stabilisation of the initiative: new intepretations, attitudes and tensions

Shortly after it had been launched officially, Digital City Bristol was already in the
need of a re-definition, as the power of novelty, together with the initial targets set
by some of its shapers, had started wearing off. All of the main actors involved
were recognising elements of success, which were of course related to their own
expectations from the project. Obviously the shared goal of facilitating partnership,
and linking some influential Bristolian organisations had been achieved.

As we have noted before, the City Council was certainly satisfied about the
image return from the operation, its coherence with the wider 'IT in the
Community' strategy, and interestingly the fact that the Council's own website,
clearly linked and advertised on Digital Bristol, had increased its visibility. The
marketing of the city and of the Council itself seemed to work fine: 'The most
frequently visited areas of DCB are in fact the city council pages, and within our
pages the tourist information and "about Bristol" pages are the most popular'
(Bristol City Council, 1997a, p.6).

Hewlett Packard Labs had been able to do some 'live' experiments with Java-
based community networking, and content distribution, enacting their 'good
neighbour' policy. UWE had started networking with the other two important
actors involved, and had carried out some prominent research and development.

However, differences and divergent visions had soon become stronger than the good reasons to work together. The cohesion of the management team, and above all the trajectory of the project and its momentum, had suffered because of these different interpretations and expectations.

As a result, despite a certain amount of self-confidence manifested in talks by the interviewees as well as in their documents and papers, Digital City Bristol's role and character were not very clear at all towards the end of 1997. And indeed tensions and dualisms were going to last throughout the two or three years to follow, if we have to judge things from de Bruine's paper presented at a conference at the end of 1999. In her paper she still refers to the tensions existing between the DDS-based paradigm supported by the HP senior researcher, and the broadcasting model that was increasingly becoming dominant: 'It is important to focus on communication in the web site to make it more than an information database' (De Bruine, 2000, p.124).

Several similar tensions had been developing between different aspects of the project. The limited resources, both human and equipment, allocated to the digital city were appropriate for a research operation with a limited scope, but definitely insufficient for a fully-fledged civic service, even if just mono-directional, as noted by the head of Council IT services:

> What happens with many of the directories on the Internet is that they are started on the wave of enthusiasm by just one bloke in his bedroom, but then keeping them up to date and working is demanding and they do not go forward. That is why DCB needs to have a proper organisation, with staff, money, etc. It has to be a mini-business. It needs to have a life of its own.

And the webmaster, who was in fact acting as the 'bloke in his bedroom' described in the previous quote, confirmed that the pressure was mounting and making the project unsustainable:

> I know for a fact that in the next few months I will be overworking. Almost everybody who is being mailed by us replies asking to be in the directory, and many want a web page and they need assistance. So the answer is no, I do not think that one person can be enough.

Another important issue that was creating problems was about the interface and functionality of the digital city. We have seen how strong the dualism between the participative paradigm and the broadcasting one had been. This was clearly reflected in the rejection by some partners of the initial concept design of DCBI, and its radical transformation that has already been mentioned in the previous sections, abandoning the DDS-like model:

> The fact that the design was derived from DDS Amsterdam has had also a potentially bad influence over some influential people in Bristol, who have made some negative comments about the design like 'it looks as if it were made for a eight year old' (UWE research coordinator, 1997).

Still connected to this, a 'browsing vs searching' dualism had been developing, about what was best for the digital city's interface. This was not just a matter of technological preference, but a dilemma deeply rooted in the different conceptions of the initiative. The HP senior researcher claimed that DCB had to be 'browsed' to make people do more than just retrieving the specific bit of information they needed, a perspective very similar to those observed in chapter three about creating 'virtual sidewalks' and forcing people to interact. At the same time the webmaster, who was working for UWE, complained exactly about the fact that too much browsing was involved, and that people had to download pages they were not interested in, before reaching the relevant information. This was of course a perfectly legitimate thing to argue, if the digital city had to be considered a 'civic database' service, but it was nevertheless totally opposite to the HP senior researcher's position, and his vision of a communicative, bi-directional, 'social' environment.

What was DCB at the end of 1997 and in the following years? The interpretations that had contributed to start it, could also contribute to terminate it, and all of the actors could let the initiative go downhill without losing out significantly, having already enjoyed a series of benefits from it.

Most of the envisaged trajectories that had motivated the partners had either reached their end or not progressed at all because of tensions in the different approaches. Bristol City Council, although willing to keep collaborating in the project, had always kept itself on the safe side by relying on its own 'corporate' web site that was linked to the digital city. If DCB had to die, this would not have too heavy a bearing on the Council's operations, and many of its initiatives could find another platform to be put onto. UWE had reached its networking aims, and did not have the capacity for letting the project grow further and in a sustainable way. The UWE research coordinator was 'not going to be upset if [the digital city] dies' and was openly considering a change of ownership for the initiative:

> This is a phase when we are starting debating about 'what to do next' after the pilot phase is ended. How do we hand this thing to the city? Do we want it to continue and what do we want to develop? This means that somebody else will have to get ownership.

HP had experimented with innovative interfaces and Java, and as the experiment was gradually diverging from the initial intentions and paradigm, the HP senior researcher himself was willing to hand the project over to someone else:

> Linda Stewart and I got the thing off the ground, but as we went along we started talking to more and more people. These people are part of the advisory committee. This committee has just met once with the view to work towards a sustainable model of the digital city. The running of the digital city in the future will have to do less and less with me, and I will tend to concentrate on research rather than practicalities.

So, Digital Bristol had really reached the end of its first phase of development. The partners had got some of the things they wanted out of it. The project was seen as having the potential to reach a critical mass of publishers and users that would make it unmanageable with the current, scarce resources. Sustainability was a key issue, and one that the current owners did not want to face on the frontline.

There seemed to be an agreement among all of the main actors on the fact that ownership of the digital city had to be handed over, to start a second phase of development of the initiative. This was mainly seen as stemming from the need to make DCB economically sustainable: 'Sustainability is the issue now. How we take this from a research project to a mature thing' (head of Council IT services, 1997), and: 'All of this does not come cheap. If I see one thing that is going to stop it from working, it is going to be money' (webmaster, 1997). Someone with money, or with the ability to turn the project into a money-making enterprise, was needed.

But the handing over process was not going to be an easy one, and in fact it would take a much longer time than expected. Beyond the agreement on the need to find new managers, clear tensions had been developing among the three partners. This looked like a consequence of several factors such as the loss of momentum and interest from the actors, the fulfilment of the initial, narrow common aims of the group, the emergence of the fragmentation of the paradigms and the shift towards a broadcasting one, and the crucial gradual withdrawal of the central technological entrepreneur, embodied by the HP senior researcher.

The head of Council IT services was explicit about these growing tensions, confirming much of what has been argued so far:

> The taking over process is going to be much more difficult than I expected at the beginning, because there are some tensions. They are organisational tensions. For example, HP's vision about this is – I guess – that people would buy more computers etc. UWE are a research academic institution, and this is from their point of view an interesting research project. It also publicises their courses and the university. In a way we could have a similar view, as we have developed quite a substantial web site that will continue even if DCB does not (...). Each of the organisations has different objectives towards the site. Eric for instance is very keen on the Java side, for obvious reasons. But we are getting complaints about Java from most of the users. So, where do we go? Do we drop the Java? HP would not be very happy about it. UWE will have new research projects coming along that could be more exciting and innovative than this, and their interest may wane overtime.

In order to discuss and plan the future of the digital city, an advisory group had been formed, including up to sixteen representatives from businesses and organisations based in Bristol. However, the effectiveness of the committee was doubtful, possibly because of the fact that the obvious fragmentation of ideas and possible trajectories was weakly managed by the core management team. A manager from Hub45, a member of the committee, in a letter sent to the UWE research coordinator lamented a lack of definition of strategy and direction for the initiative, arguing that such uncertainty would make it difficult for an external organisation like his to decide to get involved (HUB45 manager, 1997). The advisory committee was not providing a real answer to the urgent problems of DCB, being itself encumbered by fragmentation and lack of direction, as shown by De Bruine's remarks a couple of years later:

An advisory committee, which represents a range of citywide community, business, and training organisations, exists to gain valuable grassroots feedback and advice. The advisory committee is a group of around sixteen organisations from all sectors with focus on voluntary sector involvement. The advisory committee meets in theory quarterly and is chaired by a Councillor from Bristol City Council. In fact this group meets very infrequently and has not met for some time (De Bruine, 2000, p.113).

As mentioned earlier, eventually some new partners would emerge, one of them from the advisory group, to re-shape the digital city as an information service and a relatively smart directory for the city and area of Bristol. In a way was adopted what had been left in terms of a trajectory by the previous management team: a broadcasting paradigm that seemed to be the lowest common denominator for DCB. This attracted two new partners: Kaliba Netgates – formerly City Netgates – a local software house and Internet Service Provider, able to provide infrastructure and hosting, and the Bristol arm of the BBC, which confirmed that the broadcasting paradigm was becoming dominant in the digital city. In June 2001, nearly four years after the handing over process was supposed to begin, the new-look web site of Digital Bristol was launched, and with it the so-called phase two of the development. This was still supposed to be characterised by 'The introduction of more interactive elements to the website, to include virtual meeting and communication areas, multi-media and interactive spaces and dynamic news feeds' (Digital Bristol, 2001b), and the HP senior researcher was still mentioned as an active member of the management team. But observations of the site carried out in 2002 still revealed a digital city completely geared at being a hi-tech directory for Bristol-related sites. Although no further interviews could be carried out in 2001-2002, and therefore hard evidence is not available on the reasons for this, it could be argued that the many years in which the initiative had not been interactive and participative, and the consequent trend that we have mentioned, could well have contributed to the 'stabilisation' of the idea and vision behind Digital Bristol as a broadcasting, mainly mono-directional urban facility.

Conclusions

This second case study has been dealing with the history of what was launched as another supposedly advanced example of web-based digital city, developed within a different context, and through a different approach than the first case. The analysis of this case study has reviewed the arena, processes, and interpretations that were behind the electronic façade of an interesting civic web site. This has proved invaluable to understand better how this technology and its development trajectory are strictly related to factors that are essentially non-technological. It has also enriched the knowledge of the digital city phenomenon in Europe, by dealing with a Northern-European case. What main issues can we now summarise and draw from the history of Digital Bristol? The following certainly are to be considered.

Digital City Bristol underwent a very long 'pilot' phase in which there was great uncertainty in its development trajectory and feasibility. Although this limited its capacity to actively attract and retain citizens/users, the potential to grow seemed to be there, as proved by the worries for its future economic viability, and lack of resources to accommodate demand. So, DCB could have relatively quickly reached a critical mass of users if a clearer, more concerted and proactive strategy had been in place.

The presence of a partnership and a management group which did not limit the ownership of the initiative to just one actor, had the potential of letting the digital city reach different parts of Bristol's communities. Having the Council and university working together, for instance could have meant being able to address groups that often have little contact, and could have facilitated their participation. However, the balance between converging interests from the partners on the one hand, and fragmented visions and interpretations on the other, proved to be an extremely delicate – and crucial – element of the digital city's life. The common aims of the partnership were weak and short-term, and some of the very reasons that had generated the idea and initial paradigm for Digital Bristol had been never fully accepted by some of the partners, making the project quickly lose its momentum.

Leadership from the main technological entrepreneur – HP and its senior researcher – had been strong and visionary at the beginning, but as some of its main ideas were being marginalised, this had weakened, and the central role was not been taken over by anybody else. This case study seems to confirm how crucial the role – and presence – of the 'champion' of the technology is, as indicated in Guthrie and Dutton (1992, p.592).

Despite the fact that the partnership involved three very important entities within Bristol, the level of participation within those organisations was extremely low. Moreover, no professionals of the Built Environment were in the picture. No planners and economic development people, for instance, were involved on the Council side, and the digital city was being dealt with only by the IT section of the municipality. Likewise, on the UWE side, nobody from the department of the Built Environment had been included. So, the potentially beneficial effects of the partnership were greatly compromised by the lack of inclusiveness and communication within the single organisations. For the same reason, no serious perspective on the role of the digital city in relation with the planning of the real one had been developed.

As a consequence of the previous issues, the winning paradigm in Bristol ended up being what could be seen as a 'lowest common denominator' among the interpretations of the main actors: the digital city as a broadcasting entity, from organisations and institutions to people. All the two-way, community-building communication elements that had been envisaged at the dawn of Digital Bristol had been losing support, and were never properly developed, apart from some short-lived, voluntary experiments.

The only really inclusive policy to be implemented came from the Council and its initiative to establish a good number of public access points to the Internet in the city. However, this remained only loosely related to the development of the digital city, and did not have corresponding 'soft' initiatives to facilitate the establishment of a public cyberspace and encourage debate and public discourse.

PART III
ISSUES, DILEMMAS
AND THE FUTURE

Chapter 7

Lessons to be Learnt from Bologna and Bristol

Introduction

Having observed and analysed the case studies of the digital cities of Bologna and Bristol, it is now time to compare and cross-reference the two narratives in order to identify common and different results within the main themes of the cases, and try to explain these.

As Yin (1994) notes, it would be a mistake to try and generalise conclusions from a multiple case-study strategy in the same way as could be done from a survey. Cases should not be considered as a significant 'sample' of a wider population. Therefore, it is important to be aware that the results of this part of the investigation could not be generalised as 'what happened for all digital cities in Europe at the end of the 1990s'. A survey to give some answers to such a question was conducted as the first half of the strategy of the project outlined in this book. However, the insights that nevertheless came from the two case studies that were carried out, ideally complemented the survey work, providing a chance to investigate in-depth some of the underlying themes and possible explanations of the ways two very prominent digital cities were being shaped.

Yin (1994) argues further, that it should be a very similar approach to scientific experiments, where replication and non-replication of results inform and modify theories, and leave scope for further and more sophisticated experiments, rather than providing generalised evidence.

Due to the explorative nature of the whole study, as well as to the fact that the field of investigation was relatively new, little pre-existing theoretical basis was available to explain the shaping of digital cities. The survey and the case studies could contribute towards the construction of a basis to interpret and observe this new phenomenon.

This chapter is organised in sections that address the main issues that the study on the two cities raised. These issues relate to the social arena involved in constructing the digital city, and its ownership, to the relation between digital cities and strategies for urban economic development, to cyberspace as an extension of the public realm and the social access to it, and to the relation between physical and virtual town. The chapter aims to show how different interpretations and adopted paradigms affected the development of the cases considered, and their specific characteristics. It concludes by considering how processes of interpretation and social construction can be at least as important as the 'contents' of an initiative, in determining its fortunes, and its character.

Public vs private: who owned and controlled the digital city?

Although superficially it could be said that both Iperbole and Digital City Bristol were the result of a partnership among different agencies, and that in both cases some sort of 'convergence' of aims had facilitated this, comparing the two would highlight far more differences than similarities. The different socio-economic and political contingencies played a major role in the initial shaping of the two initiatives. The Italian digital city was configured as basically a Council-driven enterprise, with the other partners playing an important role on the technical side, but ending up holding a much less central position as far as the decision-making was concerned. Omega Generation had started as a proactive partner within the initial phase of design, but had gradually concentrated on the software development and its commercial exploitation with other public administrations. CINECA offered consultancies and services, without really assuming active co-ownership of the initiative. In Bologna, the role of the public administration as the sole owner of Iperbole was really never questioned by anybody. This is likely to be associated with the fact that Southern European local authorities, especially those governing major cities, tend to have considerably more power and autonomy than those in other countries, especially the UK. Although governance issues are getting more important in Italy, and more and more locally operating municipal companies are being privatised, city government still retains a great degree of prominence in urban affairs, and it is much more likely for a public administration to be perceived as 'the city' as a whole, rather than just one of the actors of a very fragmented institutional picture. This seems to be confirmed and reinforced by the more general observations drawn from the survey work on a good number of Italian cases, where single council ownership tended to be the prevalent model.

In Bristol, on the other hand, the municipality tended to follow a logic of co-operation and co-funding of the initiative that would not just involve elements of the private sector participating in decision-making, but would allow industry – in this case HP – to lead as the co-ordinating partner.

Whilst in Bologna the initial 'vision' for the digital city was mainly born within the council's walls, with a local politician and councillor setting the shape of things, in Bristol it came from an HP employee who then started 'selling' the idea to the university and the local authority. This obviously had its implications in the public/private relationship and in the way this relationship was interpreted and managed within the two digital cities.

Iperbole had been conceived as a timely alternative to any form of commodified urban cyberspace that sooner or later would have been sold in the city. It was the concretisation of a very clear political statement from the left-wing politicians and intellectuals who were opposing a growing ethos of privatisations in the country. Iperbole was intended to arouse attention well beyond the boundaries of the town. Initially at least, it was a highly symbolic initiative against the right-wing liberal politics of the government lead by the media tycoon Silvio Berlusconi. It was going to symbolise the medium of the left wing, interactive rather than mono-directional in the way of Berlusconi's television channels, free to access and liberated from advertising of all sorts, including the political ones. Moreover, Iperbole was also meant to boost political and administrative transparency within the council-citizens' relationship, a need strongly felt in bureaucracy-haunted Italy. So, Iperbole had to be almost exaggeratedly public, lead by the local authority as a means to grant universal access to the Internet to its citizens, reinforcing its social mission but also its perceived need to control the virtual environment in order to avoid any feared commercial contamination, or any chance of groups or individuals exploiting the net for their own political advantage.

Digital City Bristol had to be managed in a totally different way. The council had a much weaker influence than its Italian counterpart within the urban scene. However, it could be argued that it was itself responding – positively in this case – to the political climate and the guidelines set by the country's central government, by which the need to launch and manage initiatives by public-private partnerships was becoming an imperative. The mixed ownership of Digital Bristol meant that the initiative was not going to be run by a municipal office – as in the case of Bologna – but by a management committee including representatives from the three actors. Having been generated by the private sector, although the senior HP researcher's vision was inspired by a non-profit and inclusive enterprise such as the early Digital City Amsterdam, the Bristol experiment was much less wary of private or commercial influences on the digital city. Indeed, one of the main reasons for pursuing this partnership from all parties was openly a promotional one and, in the case of the city council, a clear emphasis on the economic development potential of urban cyberspace, rather than universal social access as in Bologna's case.

In the two cases there seemed to be a correlation, therefore, between the specific socio-political contingencies at the time of the establishment of the digital cities, and the way they were initially organised and shaped. This in a way would not come as a surprise, as it confirmed previous case studies on the construction of urban technology initiatives, such as the work by Guthrie and Dutton (1992), which gives reasons to reflect on these case studies.

In general, the observation of Iperbole and Digital City Bristol suggested the validity of the social constructivist theories and approach to technology. Despite the fact that the two initiatives were developed roughly at the same time, and by implementing very similar technological devices – primarily the World Wide Web – in two European cities, the choices behind their shapes and fortunes were dictated by political, economical, and social visions, and the related interpretations of what a city is and how it works.

The leadership factor: technological entrepreneurs and the digital city

One of the main and most interesting findings of the research carried out by Guthrie and Dutton was the crucial importance of clear and committed leadership, for the successful initiation of an urban computer-mediated communication project.

They noted that: 'In each city, a particular department emerged as a primary proponent of an information utility. Absent this champion, it was not built' (Guthrie and Dutton, 1992, p.592).

The two case studies shown in this book seem to confirm the importance of this issue, as both Iperbole and Digital City Bristol became realities through the activities of their main technological entrepreneurs, who were, at least in the initial phases of development, the constructors of the arena, by their activities in enrolling other actors and partners, as well as being the providers of the visions that would help 'sell' the initiative and give it a trajectory.

The partnership between the Council secretary for Innovation and the project manager in Bologna was particularly effective as it could address different audiences such as local politicians as well as office managers and workers of the council. The strong political vision from the Council's secretary for Innovation was also very timely and acceptable by the leaders of the Italian left wing, a fact that provided a very good level of support from the local and national press. That same vision ended up generating – and gaining acceptance of – the strong statement and policy about universal access and free Internet in Bologna, that although it would make much less sense nowadays, as many telecommunications companies are offering 'free' Internet deals, was at the time seen as a truly revolutionary move.

We have also seen how Digital Bristol was conceived and promoted by the senior researcher of Hewlett Packard, who before anything else managed to sell the idea to his own company and his managers. He actually started a preliminary study that outlined some of the main visions behind his conception of the digital city, and that were then used to draw in other actors – even though these ended up having some quite different expectations. But it can surely be argued that without the HP senior researcher, Digital Bristol would not have been born, as the council had an ambitious but definitely less 'holistic' agenda to pursue about IT and the city.

For both cases there were, of course, some further important contextual factors to be taken into account. The cultural, economic and technological 'climate' of the time could be seen as a fundamental reason for the establishment of these initiatives. The entrepreneurs were 'facilitated' by the historical moment. The 1990s, and their second half in particular, were the ideal context in which to launch an Internet-based urban initiative, and indeed, any Internet initiative. The 'hype' was high about new technologies, the media sensitive and interested in what was happening, and IT giants like HP were more than keen on investing and going for areas that definitely looked able to provide a steep and long-lasting growth. If the HP senior researcher in Bristol or the secretary for Innovation in Bologna would have tried to 'sell' the digital city idea in 2002 – at an all-time point of low confidence in the economic potential of IT enterprises – the story might well have been different.

The other factor that was playing a major role in inspiring and giving credibility to both project leaders was the emergence of other similar experiments of virtual communities based on the urban metaphor in the US and Europe. In particular, the more mature and highly celebrated digital city in Amsterdam, better known as DDS, was very well known to both project teams, and indeed had directly inspired the HP senior researcher's vision.

It is also important to notice, however, that in both cases the 'weight' of the main technological entrepreneur and of his/her visions tended to weaken or disappear with the advancement of the development process, and the emergence of other actors' visions and needs. This was evident both in Bologna and Bristol, and it showed how less visionary and less potentially innovative the initiatives were becoming as both the Bologna secretary for Innovation's and the HP senior researcher's influences lost momentum. This was particularly true for the ideas of establishing a parallel 'cyber' form of public space, which had initially driven both digital cities, although with slightly different emphases. Iperbole and Digital Bristol tended to become vehicles to deliver a relatively fragmented series of services or information. While the Italian cybercity could still benefit from a certain amount of coherence given by its 'single ownership' and the continuous presence of the same project manager, Digital Bristol found itself facing – since the end of 1997 – a definite lack of leadership, and a consequent crisis of identity and doubts over its viability, that eventually resulted in a radical change of ownership, mission, and design.

The presence of a strongly motivated and visionary technological entrepreneur seems therefore to have been crucial for the initial success of both the initiatives, regardless of whether this promoter was coming from the public or the private sector.

Local vs global: how were local economic benefits pursued?

Were the two digital cities observed pursuing clear strategies for the economic development and regeneration of their host places? What was the attitude towards business and economy in general? How were Iperbole and Digital City Bristol addressing the tension between globally oriented economic actions, and locally based ones?

Both initiatives had been born and developed with a clear awareness of their potential for the promotion of places and organisations. Indeed, one of the main functions of both experiments was promotional, although in somewhat narrow terms.

Bologna City Council was counting on Iperbole as yet another clear sign of its capacity for innovation and leadership within the clubs of European local authorities, such as Eurocities and Telecities. This was inevitably going to have a weight in the municipality's ability to win EC funds, and to be a very desirable partner for joint research and development projects. At the national level, Iperbole was a strong image-reinforcing exercise for the left wing party governing the city, in a critical moment of its life. So, Bologna's digital city was heavily marketing its creators and managers – the Council and of course the political coalition behind it – within a global 'market' of competing municipalities.

This aspect of self-promotion was very strong in Bristol as well. One of the main reasons, and indeed the one that allowed the convergence of partners towards the digital city idea, for DCB was creating a 'club' of influential organisations within the town, so that opportunities would be generated for further collaboration.

Bristol City Council was obviously interested in promoting its image nationally and with the central government, as a new and innovative local authority.

Apart from institutional promotion, place promotion was relatively strong in both cities, but not so clearly defined within the digital initiatives. Obviously these were going to contribute, and especially Bristol's policies of economic development were heavily geared towards attraction of exogenous investment, but no strategy to do so had been clearly laid out in either of the two cases. Nor was there a strategy to encourage local firms to develop, co-operate, create a larger market, or get proper visibility on the Internet.

All in all it seemed that rather than having a problem of global Vs local economic development, both Bologna and Bristol were failing to have a strategy for their digital cities as catalysts of regeneration. Bristol City Council mentioned economic regeneration as the main reason behind its IT initiatives, but nevertheless failed to introduce any clear guidelines on how DCB could contribute to this.

Different reasons could be identified for the relative lack of effectiveness of the two digital cities towards economic development. As mentioned before, Bologna had been configured and run as a very strictly public initiative. This had involved rejecting any kind of commercial presence that could compromise the 'purity' of the environment. At most the Council had been trying to involve trade organisations and – when the city manager got involved in taking decisions on Iperbole – provide better council services to businesses. As noted in the case study, the most business-friendly initiative had been the 1998 plan to establish a 'Virtual Industrial District', although this was anyway going to be just a public directory of standard information about local businesses. This attitude ended up 'overprotecting' the digital city by keeping out not just the feared media giants and big telecom corporations, but also local, young people's enterprises, which would have benefited greatly from the support of the public digital city.

A shared factor between the two cases was local competition. In Bologna this came from other information providers, some of which got to the point of suing the Council for unfair competition. The municipality had to live with this tension, and leave the commercial slice of the market to both CINECA and other private ISPs. Although this was quite coherent with the council's protective attitude towards the 'purity' of Iperbole, it widened the gap between the digital city and anything that could mean business and commerce, with possibly the only exception of the 'Million' project for local hotels.

In Bristol, although the more open attitude towards the private sector was going to facilitate collaboration with local ISPs – and in fact one of them was going to become a manager of DCB in its second phase – firms looking for Internet presence and visibility were not encouraged towards the digital city. The economic development section of the council was in fact liasing with a private IT company called SmartWest, that was providing spaces in a sort of a virtual business park for Bristol.

As it was noted in the case studies, in both cities the result of this separation between the digital city and local companies was in a way equivalent – in terms of spatial planning – of stripping the city from its businesses and relocating them in out-of-town business parks and shopping malls.

This phenomenon could be explained by – yet again – considering the interpretations of the initiatives given by the several actors involved, as Bologna was to be seen as a democratic forum where equality could not be compromised by commercial activity, and Bristol was more an inward looking enterprise, for research and liaison, than something that had to reach a multitude of local economic actors.

Limitations of planning policies and attitudes also strongly reflected into the digital city. A balance between public and private/commercial interests could possibly have been found, but it would have required an innovative approach towards policy making, not just the implementation of technological innovation.

The absence of a clear strategy towards using digital cities to support economic regeneration in both cities, apart from a relative lack of interest for this by some actors, seemed to reflect a dangerous but frequent deterministic attitude, creating the expectation that the deployment of an urban cyber-environment would generate by itself innovation in economic regeneration practices. Some of the interviewees in both cities were ready to claim that the web-based initiative was surely going to be beneficial for the local economy, but they were not able to explain exactly how this was going to happen.

Another point to consider was the absence from the picture of those people who should have been involved when it came to regeneration: planners.

Digital vs physical: civic networks and strategies for planning cities

The relationship between the digital city and the planning of the physical city, in both Bologna and Bristol, could be described through a history of absences rather than presences, of non-facts rather than facts.

This is because, despite an interesting theoretical perspective on the interplay between virtual and physical spaces, and cities and telecommunications, which was already being developed at the time of the interviews by scholars like William Mitchell or Stephen Graham, among others, city planners seemed almost totally uninterested in it.

It is a history of absences from the process of shaping the digital city itself, as in both towns no planners were even marginally involved in taking any decisions on what Iperbole or DCB should be like, and what functions they should be carrying out. Indeed, in Bristol not even the planning academics of UWE ended up actively contributing to the digital city, despite the fact that their university was one of the main players in the enterprise.

Both initiatives had, at least initially, been conceived as (cyber)spaces that would allow communication and participation. They were also meant to be 'holistic', as most layers of urban information, services, and communities were intended to be included and represented. This could have suggested using cyberspace as a planning element, an intelligent layer to juxtapose to the physical city to 'augment' or enhance some of its functionality.

But among planners the paradigm that prevailed was the 'broadcasting' one. Both digital cities were seen as a vehicle to deliver information from planners to citizens, mainly to market what was being done. A few planners could see some value in an interpretation of the digital city as a 'tool' that could be used instead of brochures and leaflets, rather than as an actual part of the city itself, that could have played a role in the actual town planning.

The most innovative approach towards the digital city that could be observed in the case studies was the senior planner's ideas of setting up an online interactive GIS for environmental information within Bologna. This was surely an interesting initiative, but still stemming from the vision of the cybertown as a sophisticated planning-aid tool.

Another way to understanding and explain the absence of planners is to consider the criticism raised in the interviews towards planners for trying to keep participation mechanisms off their area of work. As the digital city could certainly be seen as a possible vehicle for participation, this could end up being an additional reason of its unpopularity among some planners, who considered public participation as a nuisance and an obstacle to their work and professionalism.

For both cases it could be argued that the lack of engagement of town planners with information society related themes, including concrete examples of the digital city, derived from a reactive rather than proactive attitude towards managing the city, and a deterministic way of looking at the development of new technologies as a linear process and trajectory where planners stayed at the receiving end.

The interviews with planners in both Bologna and Bristol indicated that most if not all planners saw themselves as people who put things right when they go wrong, rather than designers, in a wide sense, of the future. This meant serious limitations in thinking strategically. Most planners would approach their tasks in a very traditional way, concentrating on one 'fragment' of the equation rather than considering all aspects of urban society as well as the possible interplay between space and cyberspace.

All of this demonstrated once again how a cultural and attitude shift was needed to address these issues and exploit the possibilities that could come with the new technologies, and how the mere implementation of the technology was nearly useless without this shift. A stronger emphasis on strategic planning, for instance, would have probably helped towards a more active approach with both Iperbole and DCB. The characteristics and the limits of the physical city and its planning practices, rather than being innovated by cyberspace, hampered its effectiveness and were somehow mirrored in it.

Citizens, consumers and elites: how was the right to access cyberspace interpreted in the digital city?

Iperbole started its operation with a clear and provocative statement from its creators about the role of the city council being that of serving citizens and not 'clients'. There was a clear will to highlight the differences in interpretations of the users of council services between Northern European, more liberal forms of government, and the Bolognese social democracy.

This, together with the already mentioned wariness for any kind of commercial influence within the initiative, kept the conceptions of 'users' of the digital city and 'consumers' well apart and somehow incompatible. The digital city had to be used, not consumed. It was an environment, not a commodity.

Access to the Internet world was seen by Bologna's secretary for Innovation as a democratic right of all citizens, something the public sector should have been providing universally, the very same way roads or public lights had to be provided and maintained. Access would have been the key to encourage more citizens to get acquainted with the net, and improve their computer literacy and ability to deal with the new frontier of cyberspace. This opened the way for what was seen in 1996 as truly revolutionary policy making: offering free full-Internet connections to all citizens. In the short term such a measure worked out to be extremely successful for Iperbole, as membership grew rapidly and beyond expectations. It could be argued that Bologna was certainly one of the most cyberspace friendly and aware cities in Europe, if not the most.

This attitude however changed radically when other actors, like the city manager, started getting involved – and the (former) secretary for Innovation gradually disappeared from the scene. The new management talked more happily about 'clients' and had a more consumer-oriented ethos. The city manager was also strongly doubtful of the worthiness of the free Internet policy, and claimed that universal access issues were not relevant to him. This was a major culture shift for Iperbole, changing its mission from being a 'cyber-place' targeted at everybody to becoming a 'cyber-service' for a series of more specific groups of council's customers.

Of course it would not have made sense to keep the free Internet provision alive for much longer, as similar offers were starting being available from telecom providers. But beyond the practicalities, the shift of emphasis and interpretation of what the role of Iperbole had to be, and who its users were was a major change. The free Internet initiative was, apart from its usefulness at the time, a powerful symbol of what the ethos of Iperbole was supposed to be. Although that initiative was becoming economically hard to sustain, and pretty out of date anyway, other more appropriate symbolic statements on the same line of thought could have followed it, but the shift in interpretation did not allow this.

It has to be acknowledged how Bologna achieved some notable results on the accessibility side thanks to strategic thinking, which coupled the mere implementation of new technologies with actual policy making at the urban government level. However, more could and should have been done in terms of implementing policies to tackle the computer literacy, gender and age barriers, as demonstrated by the results of the internal Iperbole survey. It cannot be argued though that the second, more service-oriented phase of the Italian initiative had nothing to do with access. Indeed, better and more versatile services could attract people and constitute a strong motivational element to join the civic network.

However, the shift from one interpretation to the other meant changes in the ways technology itself was going to be implemented. For instance, at the beginning of the project the need for public access points had been identified with the need of deploying publicly owned and manned PCs in libraries and public or semi-public spaces. This made perfect sense as public participation in debates, and empowerment of the citizens through access to the vast repository of information that the Internet was, meant enabling people to use a fully-fledged computer. The following phase, though, saw a partial re-definition of these expectations, and the implementation targets that came with them. Shifting the emphasis from a participative environment to the distribution of advanced urban services, meant that the public access points could be identified with already existing ATM machines, as what was needed was the ability to initiate simple transactions, in some cases involving the transfer of money. So, Iperbole's managers started concentrating on ways to liase with local banks in order to enhance their cash machines with civic services. Needless to say, a cash machine is not a suitable interface for exchanging ideas, finding jobs, or learning about a voluntary group. The idea of public PCs was not abandoned altogether in Bologna, but the expectations about this were definitely lowered, and the intervention was heavily downsized.

In Bristol there was a strong interest from the Council towards granting access to disadvantaged groups and in general achieving universal coverage for the access to the network. A very similar conception of the importance of urban cyberspace to that of the secretary for Innovation in Bologna was well stated in Bristol's IT agenda, where IT infrastructure was seen as becoming as crucial as roads and buildings (Bristol City Council, 1997a, p.6). Whilst Bologna City Council was trying to actively include and grow 'home' users through its free Internet policy first and the subsidised training later, its Bristol counterpart seemed to be trying to boost usage and hopefully literacy through a wide provision of public computers.

Again, lack of an overall, holistic strategy towards IT and the city, and IT and the citizens, meant that in Bristol the Council ended up being very active on the public access terminals side, but not active at all in other important aspects of the relationship between cyberspace and the city. It happened in a different way than in Bologna, where the municipality had overall control over the 'soft' side of the cybercity. It has also been noted how Bristol was a much younger and less prominent local authority than Bologna, within both national and international arenas, and that this meant that Bristol had less ability to win external funding, especially from the European Community.

This concern for universal access was however not mirrored by the interest of other partners in the Bristol case, and their way to interpret the initiative. UWE in particular had been considering the digital city as a research experiment, and a tool to reinforce a network of influential institutions in Bristol. This meant that from the point of view of the university, it would have made very little sense to try and include everybody, as all that was needed was an interesting 'sample' of participants for the research, and a very specific and low number of significant institutions for the networking exercise.

This more 'elitist' view of the cybertown arena, coupled with the financial constraints of a situation in which none of the partners was ready or able to invest large sums of money, opened the way to a convergence of visions towards a more mono-directional digital city, where selected groups could inform the mass. This 'broadcasting' paradigm – present in many other initiatives as noted by the survey work – was in Bristol a sort of minimum common denominator allowing the partnership to work together, and head towards handing the initiative over to other actors from the private sector.

In both initiatives an emphasis was put on organised groups rather than individuals, giving the former more rights and privileges. In Bologna in particular, web space in the digital city was available only to recognised associations, and their ability to update information was strictly controlled, an attitude reflecting the fact that in Italian cities the emphasis was very much still on urban government, rather than governance, and this was mirrored in the Council-driven cybertown. Bologna City Council was keeping a very firm editorial control over the electronic city, as the single owner and manager of it. However, even in the Northern European Bristol, control over the digital city contents and inhabitants was strong, as the leading institutions tended to select those groups and associations that were seen as functional to the project, and appropriate for the development of Digital Bristol. Concern about potential damage to the image of the owners of the initiative was also a driving factor for controlling access and contents.

So, a tension was apparent in both cases, between the desire to open access and allow an increasing number of citizens and organisations to use the digital city facilities, and the need to keep 'order', recognisable meanings, and above all retain control over urban cyberspace. The case studies confirmed how important it is to consider differential levels of access and ability to engage in activities, as also explored in the previous chapters.

Active vs passive access: how were the leading paradigms and trajectories affecting public discourse?

CityCard/Iperbole started with a clear and very strong emphasis towards the construction of a public cyberspace for the citizens to share and use actively. The view from the secretary for Innovation and the other main actors involved was centred on two-way communication between the citizens, as well as between citizens and the public administration. Discussion areas as well as facilities to email the Mayor and many council offices were put in place early in the project, and kept alive throughout it. Beside these, over 400 voluntary and non-profit local organisations had been drawn into the digital city, and had been allowed to build their own web-based presence. The democratising potential of the initiative was explicitly recognised and fostered by its managers as well as the very friendly press commenting on it.

Similarly in Bristol the initial intention of the main technological entrepreneur had been towards the establishment of a digital city as a communicative place, a people-to-people arena similar to the paradigmatic solution that had inspired it in the first place: Amsterdam's DDS. The early efforts of the senior researcher and its HP-funded team to define the digital counterpart of Bristol were geared towards designing public cyberspace and some tools to allow its inhabitants to communicate.

However, as we have seen in the case studies, the strong emphasis towards public participation and public discourse weakened in both digital cities, to the point of disappearing almost completely in Bristol. We have also seen that in Bologna, even though more had been done to facilitate bi-directional communication and exchange, the impacts of this on the enhancement of public participation in the city had been negligible.

Interestingly, a comparison of these two rather different initiatives points at some factors that were in common, with some obvious slight discrepancies, to determine the lack of effectiveness in enhancing public discourse in both digital cities.

A main, overall, factor could surely be recognised as – yet again – a deterministic expectancy that the introduction of digital technologies at the urban level would necessarily induce a change for the better.

In Bologna, the secretary for Innovation was convinced that the deployment of Iperbole's communication facilities would boost governance processes, self-representation, and participation in an otherwise increasingly unmanageable city. In Bristol Long was foreseeing the enhancement of communication between local government and citizens, thanks to the provision of email accounts to Councillors.

In both cities, the expectations were high, and based on a 'linear' conception of the relationship between technological advancements and the urban environment: technology was an 'independent' exogenous variable, that was going to have an impact – and change – the 'dependent' city. What happened in both cases, instead, was somehow circular, or at least bi-directional. Technology might have introduced some changes to the city, but it was also evident that the city, its problems, its institutions and the way these worked and thought, impacted heavily on technological implementation, and changed its trajectory significantly.

So, in both cities Council officials and several institutions tended to resist the innovation, or kept applying old working practices and paradigms to the new tools, such as treating emails the same way as normal letters, and allowing up to 30 days to answer them.

In Bologna, in the wider urban scene, increased public participation apathy was not tackled proactively within neighbourhoods, with non-digital strategies, and the 'technological fix' was expected to work by itself. Evidence shows that it did not, and was rather negatively affected by the more general lack of participation, instead of changing it. This also reinforced a sort of a vicious circle of interpretation of the efficiency of the digital city. Being seen by Council officials and local politicians as inadequate and not representative enough, these people would not support it or use it properly, thus making it even less effective.

Another hurdle still very much linked to old practices and visions affecting new technologies was in both cases the 'sanitation' of the public cyber-arena, for fear of adverse political reactions or 'embarrassment' that could have resulted from people speaking freely within public cyberspace. If on the one hand, in fact, a vision of governance and openness seemed to characterise the ethos of the early design of these digital cities, government and strict control was still going to prevail in the intentions and actions not just of Bologna City Council, but of the elitist public-private management group in Bristol, too. This was reinforced by the slow marginalisation of those technological entrepreneurs who had started the projects.

Soon, the city was not seen any longer as a complex and fragmented organism that could have been managed only by acquiring the local knowledge and creativity of its fragments, therefore involving communities heavily in the running of the city. It was rather seen as a machine to fine tune, and an audience ready to consume information and services, rather than taking decisions or contributing to them. In a way this happened too soon, especially in Bristol, where the shift from the DDS-like city imagined by the HP senior researcher to the more controlled and mono-directional environment preferred by the other partners, did not allow enough time and resources for any truly bi-directional communication facility to be developed and deployed in the digital city. So, no public discourse friendly configuration could stabilise at all in Bristol, and the momentum of the initial stages did not leave a concrete legacy.

The 'broadcasting' paradigm became quickly the suitable and successful one in Bristol, while Bologna was re-configuring itself as an advanced urban service and information site, still retaining its participative facilities, but being able to allocate not enough resources and strategic thinking to make them really work as they had been intended.

How visions and paradigms really made a difference

Analysing and comparing the early history of the shaping of two leading examples of the digital city phenomenon in Europe lead to a series of reflections about the crucial importance of the processes that laid behind what could be directly observed about these initiatives in terms of their contents and officially declared intentions.

Visions, it can be argued, were extremely important. And by 'visions' it should not just be meant ways to look at technology. How the several actors involved in the shaping of these 'digital cities' were interpreting the nature of the physical city, the role of government and governance within the urban arena, the meaning of community, and their own role within this context, was making all the difference.

It was noted how important had been the presence of a strongly motivated and empowered main technological entrepreneur, and how one or more initial exemplar technological solutions could inspire the early design of the initiative. But despite this, widespread tendencies to pigeonholing urban functions – and therefore actors themselves – had affected negatively the potential of the digital cities considered to produce concrete effects in the physical town. On the one hand there was a strong culture within municipal organisations reinforcing a clear divide between computer people and matters, and planning issues, even if on the other hand the problems and issues of the physical city were very much mirrored and relevant in the digital city, weakening its potential for innovation in economic and community development.

In both cases there was a clear and strong tension between a vision of the city as a fragmented, organic, complex entity or as a machine to control and improve.

Different needs were associated to these visions, leading towards a more anarchic and participative system, or to a more controlled set of advanced urban services and broadcasting facility for an empowered political and institutional elite.

Paradigms were adopted, the prevailing ones being a service facility in Bologna and an Internet-based broadcasting station in Bristol, and the relevant urban technology and policies were deployed and run accordingly. A summary of these issues and their relation with the evolution of the two projects is provided in Table 7.1.

Table 7.1 Case studies comparison summary

Iperbole – establishment

Strong role of 'champion'
Complex, postmodern city
Need for communication and participation
Urban cyberspace as a basic civic right and public service
Public sector at the helm
Institutional promotion
Political statetement

DCBI - establishment

Strong role of 'champion'
Complex but disconnected city
Need for communication and high interactivity
Strong influence of DDS
Institutional promotion
Institutional networking
R&D opportunity

Iperbole - evolution

No 'champion'
City of 'clients' and economic subjects
Need for effective management and services
Urban cyberspace as a vehicle for service delivery
Participation secondary, and as consequence of good delivery of services

DCB – evolution

No 'champion'
City of institutions and citizens as audience
Need for effective broadcasting and universal information delivery
Institutional promotion

	Iperbole	DCB
Economic Development	Place promotion. Strict public ethos. Only trade organisations involved. Businesses kept out	Place promotion. Businesses involvement envisaged, but no strategy. Council supports parallel initiative
Relation with physical city	No involvement of planners, and no strategies for articulating this. Potential for online planning information	Symbolic. Expressed in the interface, but no involvement of planners and no strategies for articulating this.
Access and users	Universal subsidised access. Training for content providers, but strong control on contents. Limited effort on public access points	Open to residents, but no subsidies. Council's own effort for public access terminals
Support for public discourse	Discussion facilities as strong point, but no parallel, physical initiatives and overall strategies. Deterministic hopes.	People to people arena ethos, but facilities never developed properly. Deterministic hopes

	Iperbole	DCB
Economic Development	More open to businesses but still pursuing centralised governed solutions	Open to business links but no overall strategies for economic development
Relation with physical city	Planning and GIS information implemented online, but still no strategies for physical/cyber articulation	No strategic planning and planners involved. Shifts further from city and gets more globally oriented
Access and users	No subsidised access. Users as recipient of services. Cheap training available on basic Internet. ATMs as public terminals	Increasing emphasis on selected groups and institutions as privileged users. General population as 'audience'
Support for public discourse	Changed emphasis towards services. Participation facilities still present, but in background	Broadcasting emphasis takes over. No participation facilities.

As mentioned at the beginning of this chapter, these observations as well as the reflections associated with them cannot be automatically generalised, and they should be regarded – in a way – as results of an experiment. As an experiment, however, they suggest possibly important issues to consider for the development of a wider theory about the shaping and the deployment of urban ICT initiatives.

Inevitably, these experiments call for further experiments and research, whilst at the same time can provide suggestions for those policy makers and technological entrepreneurs willing to innovate or improve their own urban public information and communication systems, or simply reflect about them. Indeed, application of the recommendations that can be generated by this study will be in itself cause for further experimentation on the field, and will allow – if followed by an appropriate and rigorous analysis – further enrichment and refinement of the theory.

The next and final chapter is dedicated to drawing everything together into a critical picture of the European digital city phenomenon at the turn of the century, proposing recommendations and summarising dilemmas for the construction and implementation of these initiatives and beyond, towards the projects of our future 'digital' cities.

Chapter 8

Conclusions:
What Next for the Digital City?

A multi-faceted topic

In the mid 1990s, the booming technological development of the Internet and the World Wide Web seemed unstoppable and destined to grow exponentially. Together with the fast deployment of new products for information and communication, both in terms of hardware and software, a massive amount of hype was generated in the press, the media, and in scholarly papers as well, on the forecast revolutionary and beneficial impacts of the Internet on Western societies. This had also encouraged visions of urban futures in which information and communication technologies were going to be central to city management and the exercise of local democracy.

This book has considered one of the most notable urban technological developments born and emerged in the 1990s – digital cities – in the European context. The overall approach of this analysis was on doing 'real world' research rather than indulging on forecasting futures, in looking at what was happening 'there and then' and going beyond the hype that was dominating the debate on urban new technologies in the 90s. Moreover, another significant characteristic of this work was not just considering the contents, and what was immediately observable, of the digital city phenomenon, but also going beyond these, and looking at the processes, rather than just the products, underlying the emergence of these initiatives.

Such a composite approach was very appropriate to address the complex question of trying to understand how European cities were developing their own virtual spaces on the Internet, and what consequences this could have for the 'augmentation' of the physical city by its digital counterpart. This major question was then addressed through a series of sub-questions that identified a clear need to deal with contents and processes, implementation as well as social construction, visions, interpretations.

The initial survey work proved a valuable way to observe and analyse the 'surface' of many city-related web sites, and understand what the state of the art was in the development of such initiatives all around the EU countries. But above all it was extremely significant in making sense of the whole 'movement', orienteering in what was very much uncharted territory, and creating a content-based typology of these initiatives.

The case study work was inevitably much more focused, less generalisable but nonetheless very enlightening. Case studies provided an insight on what lay beyond the contents and the gloss of the web pages. A critical history of two very relevant initiatives was built and analysed on the basis of a social constructivist perspective, and this generated reflections and ideas that could definitely offer contribution to theory-making on cities and high technologies, as well as a deeper awareness to those practitioners involved in creating and managing digital cities, who often are far too busy to look beyond the immediate problems of deploying technologies and initiatives.

The purpose of this chapter is to put together results and reflections from the study, not just to summarise what was achieved, but to suggest possible ideas for improving the economic and above all community development potential of urban telematics initiatives, in the light of the findings of this research. These conclusions also highlight how some of the major dilemmas for the digital city are still valid, and in a way more relevant than ever, at the beginning of the new millennium.

How civic Internet sites were implemented in the EU

The survey showed how the digital city phenomenon within the EU was much less homogeneous than might have appeared through the over-enthusiastic contemporary comments on the forthcoming civic cyberspace that could be found in newspapers, magazines, television documentaries, as well as the some of the promotional information produced by organisations like Telecities or European Digital Cities.

A wide range of experiments was observed, and indeed the fact that a vast majority of these could be identified as having the sole purpose of being a 'global' brochure, or at most a static and mono-directional database, helped recognise how implementing a civic web site and establishing some local, public form of cyberspace could not be considered to be the same thing. The fact that many web sites in Europe were proactive in exploiting the urban metaphor to present their information, and called themselves a digital city or a digital community, did not mean that initiatives of this kind were necessarily and thoroughly geared at local communities, citizens, and urban regeneration.

In most cases, the relationship between the electronic and the physical city was very loose, and the 'impacts' of those kinds of urban cyberspace on the 'real' environment were basically non-existent.

A typology of initiatives was established, and existing experiments were classified accordingly. From this, a group of just over twenty initiatives could be identified as the best practice 'model' of development of a digital city in the late 1990s, as they tried to address, through their contents and interfaces, the needs to provide original, wide-ranging local information and accessible participatory cyber-spaces, as well as retain some local emphasis, with contents geared at local citizens and markets. These were defined as 'holistic', or 'embedded' digital cities.

The in-the-field, qualitative investigation of two of these 'embedded' examples derived from the survey, lead to an in-depth consideration of those organisational and social aspects of the construction of digital cities that could not be observed from a distance.

Through all of this a series of conclusions and recommendations for best practice in developing and managing digital cities could be drawn.

In the eye of the beholder: ownership of digital cities, and the need to be aware of the visions informing their shaping

Throughout this book it was highlighted how crucial different visions about the city and its government and governance were to determine the shaping of high technology urban initiatives like the digital cities. These however could not be linked directly and deterministically to certain types of ownership of the technologies.

The survey showed that in fact the presence of advanced experiments of the 'embedded' type could not be related to either single ownership from the public sector, or forms of collaboration through public-private partnerships, or indeed sole private ownership.

The study suggested how different ways of owning and managing the initiatives were indeed a consequence of the socio-political climate of different countries and realities, and the consequent different attitudes towards city government and governance. But it seemed that good practice could not be simplistically linked to distinctions between 'public', 'public-private', and 'private'. The case studies showed how important it was to look beyond those labels, and find out how post-modern, governance-keen attitudes and operations could be more theoretical than actual within a public-private scenario like the one of Bristol digital city.

What seemed really to matter was the interpretation that different technological entrepreneurs would make of the city itself, its management, and the role of digital technologies within this context, regardless of whether these entrepreneurs were belonging to the public or private sector.

So, the augmented city with its physical and digital layers, could be seen as a machine-like entity in desperate need of central expert control, endowed with an audience of citizens in desperate need of efficient information broadcasting, or a complex, nearly unmanageable organism of relatively autonomous and intelligent fragments that needed to be the real city 'managers', as well as its true repository of expert knowledge. The city, through its advanced technology, could be seen as the ultimate 'control-freak' dream, where the digital town would restore the ability to centrally control and govern urban society, or it could be the sublimation of a post-modern vision of wisdom and governance of the fragments, enabling community groups and individuals to self-represent themselves in the liveliest arena ever available: urban cyberspace.

Obviously, it could not be among the aims of this approach to try to recommend one specific 'vision' instead of another. It would be arrogant to suggest a one-size-fits-all solution, or even more to expect everybody, everywhere, to have the same 'right' vision of things. What can be recommended, though, is that a constant awareness of what perspective is being chosen and implemented would be crucial. Aims, strategies, and developments should be closely compared, and actions taken if these do not match. This involves research and analysis to run in parallel with, and inform the design and deployment of the initiatives. But this is something very hard to enrol entrepreneurs on, and that indeed would probably be feared, rather than encouraged, in most cases. After all, the rigorous and thorough checking of initiatives which are often relatively short-lived, and reliant on external funding, is seldom seem as a desirable activity to spend money and time on. The very low participation of academic researchers and social scientists in practitioners dominated meetings like the Telecities conferences and workshops, and the fact that this participation has never been very proactively encouraged by these organisations, highlights the problem of the gap between practice and social analysis. This is as true in 2005 as it was in the 1990s, and the results of this research seem to suggest that a closer collaboration between university researchers and technological entrepreneurs could give the digital city initiatives a deeper self-awareness, ultimately making their shaping more coherent and effective.

The trouble with determinism

Collaborative research and reflection could have helped creators and managers of the initiatives to be less mislead by the deterministic assumptions that just implementing technology would inevitably set a precise, immutable beneficial trajectory for the city.

Technological determinism could – and still can – be recognised as a factor causing digital city entrepreneurs to concentrate solely on the adoption and deployment of specific parts of the initiatives, and expecting technological development to become an unstoppable force driving change by itself. So, embedded, holistic digital cities tended to be 'planned' only in terms of the quantity and quality of their contents and, in the best examples, the ability for a large amount of people to access these. As has been noted above for the visions and interpretations of the city, what was nearly if not totally absent was reflection or research on the processes that were being put in place. The emphasis was put on the final product and its features, without considering carefully and deeply enough the many possible ways to get to those results, in terms of who should be involved and how, and what problems of the physical city were likely to be replicated, if not amplified, in the virtual one.

Indeed, rather than being either an effective control device for a supposedly homogenous city, or an innovative and efficient arena facilitating the dialogue among many urban fragments, the digital cities observed tended to mirror many of the problems and limitations showed by their physical counterparts. It seems that if technology can affect urban society, it is also very true that society can affect technology within a circular relationship rather than being at the end of a linear trajectory. This also meant that in Bologna a technological fix to the need for governance – envisaged by its original designers – could not materialise, as the Council's ethos towards city management – and so the digital city itself – would still be rather exclusive and government-centred.

But even the apparently more open public-private partnership in Bristol failed to really boost and enlarge the public arena, and involve many fragments of the city in the running and shaping of its digital counterpart. This, rather than a technological type of determinism, could be related to a political one. That is, the assumption, widely promoted by successive central governments in Britain, that public-private concerted action would invariably be a recipe for success and sustainability. The Bristol case study proved that the processes and choices behind – and beyond – the 'format' were crucial, and that a more composite and diverse type of management structure could not ensure 'per se' the enhancement of governance and participation in Digital City Bristol.

Planning and the digital city

What could avoiding a technological deterministic attitude mean in terms of the planning of the digital city, as well as the relationship of the digital city with the planning of the physical one?

From the research findings outlined in this book, it can be said that the nature of the digital city initiative as a process, rather than a product or a tool, should not have been overlooked by those responsible for the physical planning of Bologna and Bristol. In other words, it could be noted that instead of interpreting the cybercity as a 'device', an innovative policy-enhancing mechanism that was going to impact on the existing city, its shaping would have benefited considerably by identifying it as a complex part of the city itself. As shown by its ability to mirror existing problems and interpretations, the digital city really was and is a part of the city, not something 'above' it.

This highlights the extreme importance for cities and technological entrepreneurs of fostering a holistic, strategic view for the relationship between the city and its new technological artefacts, especially those based on IT. As already noted, urban technological innovation is not just about the mere adoption of technologies, in terms of what to implement and what not to. Digital cities might have been good for city planning, but they seemed to require careful planning themselves that would not be disjointed from the one for the physical city. Policy-making for the digital city and for the physical one should not have been interpreted as two completely different exercises.

Some issues could be recognised as strategically crucial for the planning of digital city initiatives. A first issue to be mentioned here is the need for a concerted set of policies, not all necessarily directly IT related. This research showed that just implementing a piece of technology, however sophisticated this could be, could not significantly affect or boost – for instance – economic regeneration or public participation.

Rather than being seen as two different strategies, 'digital' and spatial planning could have been seen as intertwining aspects – or methods – of the same overall strategic planning approach. This could also mean not just employing digital means towards physical 'results', but the converse as well. None of the digital cities observed in the study, either in the survey or the case studies, seemed to be supported by or associated to any 'physical' initiatives, apart from the establishment of a certain number of 'public terminal' points in the city. In general a very weak connection – or no connection at all – was observed in both case studies between strategies of spatial planning, economic regeneration, and public participation, and the potential of the digital city.

But how could this have been otherwise, considering how restricted the 'arena' of digital city decision makers had been? This is the second issue to consider, which is obviously closely connected to the previous point.

A factor for more successful and 'embedded' digital cites, seems therefore to be the presence of efforts to construct an arena as complex and diverse as possible, and in particular the importance of involving those built environment specialists – architects, planners, urban designers, urban economists – who could really build meaningful links between the physical and the virtual city.

The practice of 'pigeonholing' competences and activities, widespread within most organisations and municipalities, and observed within this study, seemed to be a serious limit for the effectiveness of urban initiatives. This seemed particularly true in the case of IT related experiments and policies, where a good amount of conservatism made it possible for groups and built environment professionals not traditionally associated with computers and networks, to consider the initiative irrelevant to them and their own strategies.

This lack of involvement was strongly evident within the case studies considered. Digital cities were interpreted mainly as something for computer scientists or at most for media specialists, and were therefore going to evolve exactly that way, as broadcasting 'Internet stations' with a strong urban flavour but a loose link to city planning and economic and community regeneration strategies.

Boosting communities? Entering the digital city and participating in it

These limitations to the decision-making arena were evident and crucial when it came to making the digital cities really participative, and participated in, so that such initiatives could be seen as facilitating public discourse, and expanding the public sphere of adopting cities.

The survey indicated how limited was the percentage of web-based urban information systems actively trying to implement some two-way form of communication. The presence of discussion areas – or newsgroups – allowing citizens to debate on relevant local issues, was especially rare in European digital cities. Moreover, even in those cases where some concrete steps had been taken to implement such facilities, the results in terms of increased participation, had often been disappointing, as the Bologna case study – together with observations of other relevant initiatives (Aurigi, 2000), seem to confirm. Things have not moved on significantly, yet.

So, setting up and managing 'public' cyber-spaces where discussion is possible, although a highly desirable thing to do, proved to be a necessary but not sufficient measure to actually support public discourse. Again, adoption and deployment of technology per se could not ensure that a certain, beneficial trajectory of community regeneration and public participation could be followed.

Interpretations of the digital city informed by traditional visions of power, government, elitism and the need for top-down control of the urban resource were still too rooted in the cases considered to allow for a much-needed, relaxed approach towards ownership, ability to publish information, and participative design of the digital city. Users were therefore seen and 'constructed' as mainly end-users of an information service. There was a growing tendency to see them as passive, empowered mainly to retrieve information. Even when local organisations were allowed and facilitated to publish their own information, a certain degree of editorial control on the one hand, and the exclusion of them from taking part in any decision-making on the shape of the initiative as a whole on the other, kept the arena of the digital city protagonists very restricted, and definitely not as diverse as it could be envisaged.

The expectation of creating lively public space within an environment much more highly controlled than that of the already battered physical space of the city was hard – or impossible at all – to come true, despite the alleged potential of new technologies. The political and managerial climate around digital cities – after the bold and innovative visions of the initial entrepreneurs had been fading out – seemed to be much more preoccupied about the government – rather than the governance – of cyberspace.

So, because of these limitations on the number and type of actors involved in creating and running the initiative, digital cities seemed to be more inspired by a simplistically modernist paradigm of the provision of homogeneous, ordered, good for everybody, urban (cyber)spaces, than by a postmodernist vision of an environment of autonomous, intelligent, empowered fragments that are made sense of, and can collaborate, through the power of the new technologies.

For these same reasons, digital cities were very much a 'pushed' technology – something that was not being asked for by citizens, and not necessarily that way – rather than an innovation 'pulled' by grassroots' needs.

It could have been different if participative, inclusive processes would have been set up to let community groups take active – if not proactive – part in the design and shaping of the digital city, empowering some of their members to be incisive and able to contribute to decision-making. This could have also, very likely, called for the need for a more flexible way of designing the cybertown, a way of allowing growth but above all customisation and a degree of spontaneity. If physical urban space often carries the signs of the communities, cultures, ethnicities that inhabit it, the same was not really true for the pigeonholed, tidy and sanitised urban cyberspaces discussed in this book.

Is the digital city still relevant?

Although this book provides some answers to its initial questions, by filling some of the knowledge gaps on a phenomenon that had been object of very little investigation, the need for a better understanding of urban ICT initiatives is certainly far from exhausted. Any research work on the one hand can produce results, answers, and sometimes 'solutions' to problems and dilemmas, but on the other hand surely generates new challenges and new needs to expand knowledge, as well as the need to constantly test and validate what has already been defined.

It could be said that this was particularly true for this study, as the phenomenon it dealt with was so new and little developed and explored, that a great deal of further research, as well as refining of methods to conduct it, would be definitely needed.

Also, things change, and when it comes to high technologies they are supposed to change at a speed which is almost impossible to keep up with. This is actually more a myth than reality, and whilst programming languages, applications and techniques might have improved relatively fast in the past few years, the overall philosophy behind digital city interventions has not evolved that deeply. However, significant changes that can partly redraw our scenario have occurred, such as a relative reduction of connectivity costs, much higher available bandwidth and the emergence of wireless/mobile computing.

Talking of interpretations of technologies, the attitude from private companies, investors, policy-makers, and the general public towards IT and telecommunications innovation has changed dramatically since the so-called 'technology bubble burst' of the beginning of the millennium. Much less hype and irrational expectations from technology characterise the very historical moment when these final sections are being compiled. As Cassidy notes, an overall more down-to-earth, balanced approach towards the potential of new technologies to radically change the economy, and society, is now being employed:

> There are several reasons why the New Economy argument turned out to be flawed. One of the most basic was that it exaggerated the role that information technology plays in the economy. Despite the rapid growth of the Internet, firms still spend more money on old-fashioned capital equipment, such as drills and welding machines, than they do on computers, telephones, and other information gadgets (...). Wings have to be attached to planes; roofs have to be put on houses; airbags have to be installed in SUVs. The Internet helps with the planning and organization of such tasks, but it doesn't turn screws or lay bricks. Nor does it operate on patients or serve businessmen lunch (Cassidy, 2002, pp.318-319).

Many – if not all – high technology companies have been going through the most difficult phase of their life in the first years of the new Century, with many e-businesses born right at the end of the 1990s having already gone bankrupt and disappeared.

This climate inevitably is going to affect interpretations of technologies, their potential and viability, and the images associated to them, from policy makers, politicians, entrepreneurs and community groups. It would therefore be extremely interesting – and indeed useful – to consider in further studies how these different visions can affect the deployment and shaping of urban ICTs now, and in the near future.

Also, these societal changes will have very likely affected the viability and survival of many of the initiatives that were surveyed in initial phases of the study described here. It is extremely significant, for instance, that probably the most famous example of digital city in Europe, Amsterdam's DDS, although still existing has in the past few years radically changed its characteristics and lost its paradigmatic role as an Internet-based community based on a clear urban metaphor. Lovink explains that

> By 1998-99 the free DDS facilities were available everywhere. Scores of new commercial providers and services had popped up all over the place (such as Hotmail, Geocities, and even free dial-up providers), offering the same services (often more extensive, better ones) than the DDS was able to provide. The free Internet services advertised massively and attracted a customers pool far removed from the idealistic concerns that used to inform the original Digital City. This resulted in a substantial quantitative, but more importantly qualitative, erosion of the DDS user base. Even if the absolute number of accounts had risen to reach an all time high mark of 160,000 in early 2000, an analysis of the use patterns showed that these could no longer be considered conducive to community building or even to socio-politically relevant information exchange – homepage building and upkeep for instance, no longer attracted much interest. The once so valuable website had turned into empty lots. Despite an overall growth of Internet use, the Digital City had lost its attractiveness for users (Lovink, 2004, p.376).

Is there still a role for digital cities in Europe? There might well be, although these initiatives could tend to be a much less 'hyped' phenomenon, something that on the one hand has to deal with a much more mature public, technologically speaking, and on the other has apparently lost its aura of innovation, to become yet another everyday 'object' Haythornthwaite and Wellman, in their interesting edited book *The Internet in Everyday Life*, support some of the considerations made so far and argue that

> The rapid contraction of the dot.com economy has brought down to earth the once-euphoric belief in the infinite possibility of Internet life. It is not as if the Internet disappeared. Instead, the light that dazzled overhead has become embedded in everyday things. A reality check is now underway about where the Internet fits into the ways in which people behave offline as well as online. We are moving from a world of Internet wizards to a world of ordinary people routinely using the Internet as an embedded part of their lives. It has become clear that the Internet is a very important thing, but not a special thing (Haythornthwaite and Wellman, 2002, pp.5-6).

So, we should expect to observe changes that make ICT in the city shed part of its symbolic power and role, to become an actual occurrence of contemporary urban lifestyles. Mobile telecommunications and Internet are going to strongly contribute to this shift in usage patterns, perceptions and attitudes. Agre claims that

> Computing is to become ubiquitous and invisible, industrial design is to merge with system design, and indeed the very concept of computing is to give way to concepts such as writing reports, driving to work, and keeping in touch with one's family. Computing, in short, is increasingly about the activities and relationships of real life, and the boundary between the real world and the world of computer-mediated services is steadily blurring away (Agre, 2004, p.416).

Consequently, on the one hand we can probably expect to see digital city initiatives becoming more ordinary and less 'revolutionary' – and indeed we have seen this beginning to happen somehow within this book's observations. Cybercity-style projects will somehow possibly get even less 'visible' as well-identifiable entities within the urban arena. But they would not be less relevant for our lives, less worthy of observation and analysis, or even less present. Iperbole in Bologna is still there and still works. The difference – or at least one of the differences – is that it has now become rather 'ordinary'. Newspapers have stopped celebrating it, and maybe less students are interested in writing essays and dissertations about it. Governments, however, are concentrating on these systems now more than ever, with a strong push towards the implementation of e-government projects and services, such as the *UK-Online* initiative aiming at achieving the 100% target of central and local government service e-delivery by the end of 2005.

On the other hand, urban computing is increasingly shifting from the web-based domain of information systems and place promotion sites, to being part of more and more aspects of city management, service provision, and indeed private, unplanned and relatively uncontrolled individual usage.

These are all very good reasons why further study and concentration on urban ICT projects and digital city-like initiatives should be welcomed. If digital cities are becoming more 'hidden', or fragmented, this should encourage scholars not to lose sight of them. As Stephen Graham suggests for a wider context:

> This is an important stage of development because technologies often have their biggest effects on society when they become, in a sense, invisible because they are taken for granted and assumed. In many cases this shift to invisibility is both a metaphorical and a physical one. People fail to notice technological artefacts and connections because they cease to be novel or exciting. But those technological artefacts and connections also tend to become more hidden, more miniaturised, and more embedded into the every day environment of homes, workplaces, transport systems, artefacts, and cities (Graham, 2004, p.415).

Dilemmas and challenges for the digital city

So, despite the obvious technical changes, the increases in bandwidth offered by ISPs, and the symbolic and interpretative shift of the Internet towards being something less 'special' or even notable, most of the indications provided by the early research work on the digital city remain relevant and valid. This is because even if the 'products' might have been changing, two main factors have to be considered: stabilisation and the importance of processes.

Stabilisation in the social shaping of technological artefacts has been explained in chapter 2, but it is worth remembering that whilst in its the early phases of development, a technological object or system is in a very fluid state, being shaped by the articulation or conflict of different interpretations of it, with time this tends to crystallise in an accepted form – and interpretation. Digital cities, intended as the civic websites dealt with by this book, have done exactly that, and apart form the addition of a few new services – indeed fewer than it could be expected – their basic 'shape' has remained more or less the same. Recent observations and further research, that are not fully analysed yet and could not be included in the book, seem to confirm this.

More recent research carried out on the hi-tech and Internet implementations and policies in Newcastle and Antwerp (Firmino, 2004) indicates that most of the issues outlined in this book have kept their validity and should be taken on board by practitioners and entrepreneurs who want to work on using ICT to improve the public sphere of cities.

All of this has to do with the fact that the research outlined in this book has recognised how the 'digital' city – and here the term is used with its wider meaning – instead of re-writing the rules of the civic game, and establishing new ones, really mirrors and has to cope with the vast majority of 'traditional' issues. Whilst techniques can change or improve, the underlying social, political, and economic processes shaping the 'digital' city are bound to the city as a whole, and change at a much slower pace that a few software versions.

It is interesting to notice, for instance, how the conflict of interpretations on universal access to dialup Internet connectivity, which was such a central issue in the 1990s' digital city, is back with a vengeance on wireless access. If officials in Bologna were seeing dialup Internet as a right of citizens, not different in essence to the right to have roads and other basic services, this rhetoric is being now re-presented and reinforced for the new wave of 'wi-fi' technologies. The recently established Manchester Wireless group argues that

> We believe that the data networks of the future will be as important infrastructure as roads and railways, and it is vital that these networks are not monopolised by a small few, but are run in the interests of the local community. Manchester Wireless aim to create a city-wide wireless network using freely available off-the-shelf hardware, and free, open-source software (Manchester Wireless, 2003).

And this really shows how issues identified in the early phases of the digital city history, and documented in this book, still apply and, indeed, should inform policy-making.

These reflections, together with the initial results and recommendations given in the first part of this chapter, lead to conclude by highlighting a set of 'dilemmas' which stem from the research outlined in this book. These were valid a few years ago and are still relevant challenges to consider for those practitioners and entrepreneurs who are working on any aspect of our 'digital' cities, as well as for planners who want to get more 'strategically savvy' on the chances of using ICT in regeneration.

Physical/visible vs virtual/invisible

Digital infrastructure exists, but – apart from the inconveniences caused by roadworks to install cabling – can be very hard to notice. Most of it is underground or locked in rather unexciting, anonymous buildings. All that usually can be 'seen' – and indeed noticed – is the 'terminal', the computer or kiosk standing as a threshold object/space to access the otherwise invisible network. The emergence of wireless networking and the increasing availability of small, mobile and personal 'terminals' wirelessly connectable – GPRS and 3G mobile phones, wi-fi PDAs etc – is reinforcing the dilemma of visibility *vs* invisibility of the digital 'bits' of the city. In a way, another related 'dualism' of fixity *vs* mobility, placement *vs* fluidity is being introduced by the increased sophistication and popularisation of mobile technologies.

All of this invisibility creates the very tangible problem of understanding the importance and communicating the impacts, relevance, benefits of the deployment of information and communication technologies in the urban scene. A public administration would find much easier to obtain consensus, for instance, for the construction of a series of new roads than for the extension of its metropolitan area broadband network, or for the implementation of a new electronic urban software-based system for job searching. Understanding the road system is still much more straightforward – and methods to do so are well developed, established and shared among experts – than making sense of what ICT can or cannot do for our cities.

This can therefore highlight the crucial importance of coupling policy-making and development with research, aimed at understanding and clarifying the role that the phenomenon has within the wider urban environment. Invisible networks and facilities can surely benefit from studies that would make their impacts and roles more visible and understandable.

However, very often the only way this problem of hidden facilities is tackled, is only by desperately trying to reify ICTs and make them visible through operations that can end up as more symbolic than meaningful or useful. Graham and Marvin, for instance, analysing the report on the 'Review and possible role of Teleports in Europe' by IBEX (1991), noted how efforts had to be made

> to increase the visual and physical impact of telecommunications in cities, as when prominent satellite dishes are developed to boost the image of high-tech office developments and teleports. In one case, for example, such a dish has been proposed purely for cosmetic reasons, even though no satellite facilities were actually technically required (Graham and Marvin, 1996, p.51).

This can also be related to some of the findings of this book, with some municipalities concentrating on the development of initiatives which were characterised by an almost exclusive vocation towards a rather standardised and unexciting practice of place marketing. These stood on the web as glorified electronic brochures of the alleged innovativeness and competitiveness of their host cities. Even the deployment of 'visible' civic kiosks for public access to Internet-based services, has sometimes been tackled more with the aim of making a statement and showing that the city was becoming 'intelligent', rather than as a well coordinated action within a more articulate strategy for encouraging inclusion and participation in the use of ICT in the city (Firmino, 2004).

Moreover, even if civic ICT can be somehow made visible through the terminal screen, this can still be seen as being basically 'virtual', operating within cyberspace and apparently not having much to do with land use, housing or spatial planning in general. Therefore it will be possibly disregarded by most urban planners as something irrelevant to their expertise. Graham and Marvin (1996, p.50) have argued that 'Urban studies and policy tend to be dominated by a concern with the visible, tangible and perceivable aspects of urban life'. This study also showed how urban planners of cities that were developing Internet-based 'digital city' initiatives, were scarcely aware, and interested, in these projects. Their attitude tended to be reactive rather than proactive towards the management of the 'digital' city, based on a deterministic way of looking at the development of new technologies as a linear process and trajectory where planners stayed at the receiving end and would intervene only if precise spatial problems would show up. The aforementioned more recent study on virtual city strategies in the cities of Newcastle and Antwerp (Firmino, 2004) seems to demonstrate that nothing significant in this respect has changed in the past few years, and most planners are interested in IT only for its potential to provide tools – such as GIS and related systems – to analyse traditional spatial problems.

Within local authorities it seems to happen that ICT-based initiatives, despite allegedly aiming at things like economic development, education, service delivery and city management, community regeneration and in general a better use of the resource-city, are not usually regarded as being part of the planners' remit. These projects are usually dealt with by information systems personnel and experts, and customer relationships offices. In substance, whilst it would seem sensible that strategic planning visions embedded this very relevant aspect of civic development and management through ICT, this rarely occurs.

A new, cross-disciplinary approach towards urban policy-making is therefore strongly needed. We have seen in Chapter 2 how for instance Koolhas and Mau have been pointing out the deficit of openness and knowledge existing within the planning profession respect to 'conceiving new modernities, partial interventions, strategic realignment' (Koolhas and Mau, 1995, p.965). Now more than ever before, a wealth of experimentation and ideas on technological interventions that will surely have a clear impact on urban management comes from computer scientists and engineers, and their R&D activities. In Kyoto, Japan, for instance, a very active group of researchers, funded by the Japan Science and Technology (JST) governmental agency, has been envisaging interfaces and solutions for the digital city of tomorrow, with apparently no active involvement or participation from planners.

It has been noted how important it is to understand and acknowledge the increasingly hybrid, 'recombinant' spatial situation in which we live (Mitchell, 1995, 1999, 2003; Horan, 2000). It is also crucial, though, to realise that to operate in it, and understand it, traditional barriers have to fall down and leave room to a more open-minded attitude towards urban planning. 'Recombinant' space can only be dealt with by a 'recombined' discipline.

This works both ways, though, and suggests the need for a 'holistic' way to conceive planning and management strategies for cities, a way that would not interpret physical and virtual as two separate dimensions, but that would encourage the interplay – and the hybridisation and interdependence – of physical and IT projects.

The book has discussed how a major limit of far too many ICT-based regeneration initiatives in Western cities has been a somehow an enthusiastically deterministic way to see impacts of IT on urban functions. Technological entrepreneurs have tended to believe that computers, networks, and software could act as a quick-fix for a variety of urban problems, by changing the rules the game was played by. This has allowed 'digital city' initiatives to be often conceived and deployed in isolation, confident that their innovative potential would be a catalyst for change. However, this approach has rarely proved effective, and it can be argued that if it is true that spatial planners should get more acquainted with 'virtual' initiatives, the same applies to the promoters of technological projects. These projects desperately need ways to relate more effectively and closely with urban spaces, established working practices and lifestyles, and all sectors of the local community.

Whole vs fragments

This 'classic' dilemma for urban space and governance is getting increasingly crucial and relevant as high technologies spread, becoming mobile and ubiquitous at the same time. The effects on urban spaces and lifestyles of the emergence of the information society do not look like something that civic administrations can make sense of, and control, very effectively. The typical development control approach to the government of urban space, rather modernist and rationalist in nature, finds itself unable to cope with a liberalised, market-driven, hard to pinpoint and highly fluid phenomenon as the utilisation of ICT within the civic arena. What can administrations do to play a significant and hopefully beneficial role in this? How can they relate with this situation in order to produce benefits and a higher quality of life for their citizens, and how can they – if at all – limit the possible drawbacks?

We have seen how in the 1990s organisations like Telecities and the civic networking movement were born with this in mind: enabling the public sector to influence positively the otherwise privately-driven world of emerging urban ICT, and basically standing as an alternative, as a way of contextualising IT, making it local, sensitive and creative towards local issues and needs (Carter, 1997). This has mainly resulted in centralised 'digital city' initiatives, frequently presented via a civic web portal, and in the best cases relatively articulate and willing to offer services and facilities that would go beyond the simple 'information' online site.

However, it has been noted how most of these projects have been – and still are – trying to cope with the whole *vs* fragments dilemma. Already in the 1990s officials in Bologna were asking themselves whether the city could be effectively managed centrally, or whether it was in itself a far too complex social environment for any central organisation to make sense of. The city was always more than the 'whole' that could represent it: it was its numerous fragments. So, a logic of 'control' of the whole, useful to easily make sense of things but limited in scope, would always be opposed to a more comprehensive but difficult to implement logic of 'participation' of the fragments. Does the digital city, meaning by this not just a website, but the virtually-enhanced city in which we live and communicate, need more control? And does it need more open participation, and above all, how can these two aspects be weighed against each other?

This is obviously connected to the more general government *vs* governance dilemma for the management of cities and the promotion of decision-making. Interestingly enough, we have seen that the enhancement of civic space and the public sphere through Internet-based initiatives, has rarely corresponded to an attempt to enhance governance, widen decision-making processes, and offer opportunities for different 'fragments' – pressure groups, spontaneous aggregations of people, marginalised sectors of the community, to communicate with each other and be granted some form of public representation. Most projects – often driven by strict economic development imperatives – have even chosen to offer an image of the city as a tidy, harmonious and rather unproblematic 'whole', an image coherent with a short-term place marketing ethos, but possibly in the need of more incisive actions towards increasing social cohesion and local governance.

Services for clients vs social rights for citizens

This is another notable tension that practitioners and decision-makers involved in conceiving and running projects for the shaping of the 'digital city' have been facing, and still have to face, and that links rather strongly with the dilemma of government Vs governance of the cybercity. Should the digital city's ethos be oriented towards the improvement of service provision, and a consequently better city management, or should it be focused on enhancing social and political links and boosting public discourse? The simple answer would obviously be that it should take care of both dimensions, by working towards the establishment of a better quality of life through both providing good, efficient services, and building a stronger sense of community. One of the Iperbole interviewees had also remarked, with a good deal of common sense, that communities do need services and that good provision of interactive, day-to-day electronically distributed services could indeed attract citizens towards engaging with the participative aspects of the online city.

However, achieving this is a difficult task that requires policy-makers, and in general those who are supposed to work on strategies for regeneration, to make constant efforts to balance their ways to interpret what the crucial role of those who live and work in the city should be. Are policies – and technological initiatives – aimed at 'clients', 'audience', 'end-users', or are they aimed at 'citizens', 'actors', 'owners' of the city? This is not just some type of moralistic distinction between more or less liberal approaches towards public administration. We have seen that it is an important and delicate choice that shapes the vision, trajectory – and indeed the effectiveness – of high-tech projects in the city, and is often overlooked in the name of the generic and urgent needs to engineer 'solutions' for problems or to make initiatives economically viable.

A very interesting aspect of this dilemma is the not easy co-existence of different modalities of use of online – or augmented – public or semi-public spaces. Service-driven environments – both physical and virtual – are rarely very social, tend to be used on an ad-hoc basis – only when they are needed – and accessed in a rather selective way, that is by dealing only with the information, people, sections or indeed spaces that are useful in certain circumstances. In Chapter 3 we have noted how pertinent was Shapiro's argument (1995) on the need for public spaces, even digital ones to be holistic and somehow inevitable, non-filterable.

So, whilst the clients Vs citizens, as well as the government Vs governance, are certainly not new issues to consider, and have not been generated specifically by the implementation of IT, the emergence of the city as a digitally enhanced space is going to reinforce this dualism rather than escaping the need to address it.

Are cities developing as 'digital' ones by means of institutions and projects managing to successfully articulate these two different ways of using urban facilities? Are ICT initiatives and strategies addressing the needs of inhabitants as both their clients and their citizens? Are service, client-oriented projects facilitating and encouraging democratic participation and social cohesion, or are they promoting a type of hi-tech individualistic city where technology mainly enables people to 'push a button' – or swipe a card – to get something? These are now more than ever some very 'hot' questions for ICT innovation in cities. Benini et al. (2003), for instance, seem to suggest that the shift of ethos from the initial pioneering phases of socially-oriented civic networking to the service-oriented project-making of most European municipalities, has so far failed to address this problem of balance.

Epilogue: the need for regeneration-centred research

Last, but possibly not least, it is worth noting the need to apply research guidelines and reflections to real-world experiences. It has been mentioned before how collaboration between academic research and actual policy-making would have benefited the initiatives by providing, perhaps uncomfortably, useful in-depth knowledge gained from a detached viewpoint. It has to be mentioned here how beneficial this collaboration could be for academic research itself, by opening up more avenues to acquire knowledge, and above all by providing a 'test-bed' for the research findings and recommendations. It would be most important to apply some of the principles and suggestions stemming from works such as this one to a series of real-world urban ICT projects, in particular those involving public participation, willing to put them into practice, in order to check and refine on the field – through further observation and research – the theoretical and practical issues at stake.

More in general, it would be necessary to add flexibility to the 'digital city' concept, as increased use of the web for the implementation of a variety of services is surely widening the scope of projects that could still fall within the cybercity category. If on the one had we can reasonably expect that urban brochure-like websites can still exist – but have surely stopped impressing anybody – many other ideas and urban technological 'fragments' have been thought and developed, adding complexity to the overall picture of the digital city phenomenon.

Indeed, this increased complexity calls now for research that can concentrate on looking at the many tasks performed by 'digital city' projects and facilities, and the articulation of these within the economy of city management and regeneration.

William Mitchell notes in his latest book on the topic how

> The trial separation of bits and atoms is now over. In the early days of the digital revolution it seemed useful to pry these elementary units of materiality and information apart (...). Now, though, the boundary between them is dissolving. Networked intelligence is being embedded everywhere, in every kind of physical system – both natural and artificial (...). Increasingly, we are living our lives at the points where electronic information flows, mobile bodies, and physical places intersect in particularly useful and engaging ways. There points are becoming the occasions for a characteristic new architecture of the twenty-first century (Mitchell, 2003, pp.3-4).

In the 1990s could be legitimate to concentrate on those relatively few examples of web implementation in major European cities. Now, however, the much higher penetration of high technologies into urban and regional environments, coupled with the increased functionality of these implementations and the availability of facilities not just from one centralised source – the 'digital city' as we have considered it in this study – but from a variety of public and commercial actors, suggest that a more task-oriented, regeneration-focused approach, rather than a web-focused one, is necessary to make sense of this increased complexity.

As noted before, many of the services and facilities that had such a strong symbolic role in the 90s are now becoming 'ordinary' aspects of our everyday lives, and our everyday cities. As Gumpert and Drucker note:

> Where does the 'digital city' exist in the scheme of things? By this time, all cities, whether by design or by accident, whether in a deteriorating or renaissance state are, to some degree, 'digital' (Gumpert and Drucker, 2003).

It seems that the 'digital' city concept is therefore gradually shifting from the early definition of civic websites and their related services, analysed in this book, to a more holistic and complex view of the city permeated by layers of information, relationships and their enabling technologies. And these technologies are often not designed with the city – and its planning and regeneration – in mind. This makes this particular study on the origins of the digital city phenomenon a useful basis to keep observing how the Internet and the World Wide Web are being deployed and used within the urban arena.

This book has itself – as the case studies it has dealt with – inevitably been influenced and characterised by the strength of a paradigm – the digital city as a holistic web-based resource – which had emerged towards the end on the century. But the 'digital city', in its wider sense, is here to stay, and most of the reflections of the book will be useful and relevant for further investigations on how urban cyberspace evolves in the future.

Bibliography

Agre P. (2004) 'Life After Cyberspace' in Graham S (ed) *The Cybercities Reader: Handbook for an Urban Digital Age*, Routledge, London

Arthur C. (1997) 'Behind the Net curtain', *The Independent*, 7-4-97

Aurigi A. (1998) 'Dentro la Città Digitale', in RUR-CENSIS, *Le Città Digitali in Italia – Rapporto 1997*, Franco Angeli, Roma

Aurigi A. (2000) 'Digital City or Urban Simulator?', in Ishida T. and Isbister K. (eds) *Digital Cities: technologies, experiences, and future perspectives*, LNCS 1765, Springer Verlag, Berlin

Aurigi A. and Graham S. (2000) 'Cyberspace and the City: The Virtual City in Europe' in Bridge G. and Watson S. (eds) *A Companion to the City*, Blackwell, Oxford

Azhar A. (1995) 'Brands on the Web', *The Guardian On-line* 15-11-95

Bannister N. (1995) 'Novelty of the Net wears off', *The Guardian On-line* 31-10-95

Bassanetti A. (1994) 'Una rete per Bologna', *Zerouno* n.154, 11/94

Batty M. (1995) 'The Computable City', Mimeo

Batty M. and Barr R. (1994) 'The Electronic Frontier, Exploring and mapping cyberspace', *Futures* n.26

Bekke H. (1991) 'Experiences and Experiments in Dutch Local Governments' in Batley R. and Stoker G. (eds) *Local Government in Europe – Trends and Developments*, Macmillan, London

Bellagamba F. and Guidi L. (1996) 'Citycard second phase – WP4: CityCard System Prototypes and Validation (III). Task 4.2 and Taks 4.3: Report on User Validation in Bologna and Wansbeck (III) – Report D4.2'

Benedikt M. (Ed) (1991) *Cyberspace First Steps*, The MIT Press, Cambridge MA

Benini M., De Cindio F., Sonnante L. (2003) 'Virtuose, a VIRTual CommUnity Open Source Engine for integrating civic networks and digital cities', paper presented at the Digital Cities 3 Workshop, Communities and Technologies conference, Amsterdam

Bianchini F. (1988) 'The Crisis of Urban Public Social Life in Britain: Origins of the Problem and Possible Responses', *Planning Practice and Research* 5

Bijker W.E. (1992) 'The Social Construction of Fluorescent Lighting, or How an Artifact Was Invented in Its Diffusion Stage', in Bijker W.E. and Law J. (eds) *Shaping Technology/Building Society: Studies in Sociotechnical Change*, MIT Press

Bijker W.E. and Law J. (1992) 'General Introduction', in Bijker W.E. and Law J. (eds) *Shaping Technology/Building Society: Studies in Sociotechnical Change*, MIT Press

Blair P. (1991) 'Trends in Local Autonomy and Democracy: Reflections from a European Perspective' in Batley R. and Stoker G. (eds) *Local Government in Europe – Trends and Developments*, Macmillan, London

Bolter J.D. and Grusin R. (1999) *Remediation: Understanding New Media*, MIT Press: Cambridge, MA

Boyer M.C. (1993) 'The city of illusion: New York's public places', in Knox P. (ed) *The Restless Urban Landscape*, Prentice Hall

Brants K., Huizenga M., and van Meerten R. (1996) 'The New Canals of Amsterdam: An Exercise in Local Electronic Democracy', *Media, Culture, and Society* 18

Bristol City Council (1995) Economic Development Strategy 1995-1998, BCC

Bristol City Council, Corporate Services Directorate, IT Division (1997a) 'IT in the Community', Mimeo

Bristol City Council (1997b) 'Information Technology in the Community', Mimeo

Brittan D. (1992) 'Being There', *Technology Review*, May/June

Brown L. (1994) 'The seven deadly sins of the information age', *Intermedia*, June/July 22(3)

Burrows R. (1995) 'Cyberpunk as a social theory', Paper presented at the BSA Annual Conference 'Contested cities', University of Leicester

Cairncross F. (1998) *The Death of Distance: How the Communication Revolution Will Change Our Lives*, Texere Publishing

Carter D. (1997) 'Digital democracy of information aristocracy?', in Loader B. (ed) *The Governance of Cyberspace*, Routledge, London

Cassidy J. (2002) *Dot.Com: The Greatest Story Ever Sold*, Allen Lane, The Penguin Press, London

Castells M. (1985) 'High Technology, Economic Restructuring, and the Urban-Regional Process in the United States', in Castells M. (ed), *High Technology, Space and Society*. Sage Publications, Beverly Hills, Calif.

Channel 4 Television (1994) 'Once upon a Time in Cyberville', transcript of programme

Channel 4 Television (1994) 'Visions of Heaven and Hell', transcript of programme

Comune di Bologna (1998a) 'Una tantum per allacciamento a Iperbole e Progetto Distribuzione Servizi',
http://www.comune.bologna.it/bologna/Bologna_cablata/unatan.htm

Comune di Bologna (1998b) 'La rete civica Iperbole/Internet',
http://www.comune.bologna.it/bologna/Bologna_cablata/iperbole.htm

Comune di Bologna (1998c) 'Progetto per un distretto economico-produttivo virtuale sulla rete civica Iperbole/Internet',
http://www.comune.bologna.it/bologna/Bologna_cablata/cv.htm

Comune di Bologna (1999a) 'Iperbole 2000 – La rete civica e il suo sviluppo',
http://www.comune.bologna.it/Iperbole2000/rete_e_sviluppo.htm

Comune di Bologna (1999b) 'Corsi di alfabetizzazione ad Internet e all'uso delle nuove tecnologie: risultati a Giungo 1999',
http://www.comune.bologna.it/risultati_alfabet.htm

Comune di Bologna (1999c) 'Campagna di alfabetizzazione per Bologna Digitale',
http://www.comune.bologna.it/alfabet2.htm

Comune di Bologna (2000a) 'Rete civica Iperbole/Internet',
http://www.comune.bologna.it/Iperbole/IPERBOLE.htm

Comune di Bologna (2000b) 'Centro di alfabetizzazione e divulgazione telematica per Bologna Digitale',
http://www.comune.bologna.it/alfabet3.htm

Comune di Bologna (2000c) 'Elenco postazioni Iperbole/Internet aperte al pubblico', mimeo, originally on Internet

Cottrill K. (1995) 'Losing its backbone', *The Guardian On-line* 30-3-95

Cottrill K. (1995) 'Poverty of resources', *The Guardian On-line* 30-3-95

Curati M. (1995) 'Messaggi dal mondo, già 6000 i visitatori telematici di Bologna', *L'Unità* 16/2/95

Davis M. (1990) *City of Quartz*, Vintage, London

Davis M. (1992) 'Beyond Blade Runner: Urban Control – The Ecology of Fear', Open Magazine Pamphlet Series, Westfield

Davoudi S. (1995) 'Dilemmas of Urban Governance', in Healey et al. (eds) *Managing Cities, The New Urban Context*, Wiley and Sons, London

De Bruine A (2000) 'Digital City Bristol: A Case Study', in Ishida T. and Isbister K. (eds) *Digital Cities: technologies, experiences, and future perspectives*, LNCS 1765, Springer Verlag, Berlin

Dear M. (1993) 'In the city, time becomes visible. Land-use planning and the emergent postmodern urbanism', mimeo, summary of oral presentation

Dear M. (1995) 'Prolegomena to a Postmodern Urbanism', in Healey et al. (eds) *Managing Cities, The New Urban Context*, Wiley and Sons, London

Digital Bristol (2001a) 'Public Access Points in Bristol', at www.digitalbristol.com

Digital Bristol (2001b) 'History and development', at www.digitalbristol.com

Dyer P. (1995) 'All wired up and nowhere to go', *The Guardian On-line* 8-6-95

Europan Commission (1996) 'Building the European Information Society for Us All. First Reflections of the High Level Group of Experts', Interim Report

Fallows J. (1985) 'The American Army and the M-16 rifle', in MacKenzie D. and Wajcman J. (eds) *The Social Shaping of Technology*, Open University Press, Bristol, PA

Firmino R. (2004) *Building the Virtual City: The Dilemmas of Integrating Strategies for Urban and Electronic Spaces*, PhD Thesis, University of Newcastle upon Tyne

Gates W. (1995) 'How technology transforms work', *The Guardian On Line*, 26-10-95

Geelhoed E. (1996) 'Comparing Digital Cities', mimeo, previously published on HP Labs website at http://www-hplb.hpl.hp.com/psl/itd/people/eg/webpubs/dcbi/comp.htm

Geelhoed E. (1996-2) 'Digital City Bristol Interactive: The Hewlett Packard Research Labs Angle', mimeo

Gelernter D. (1991) *Mirror Worlds*, Oxford University Press, Oxford

Gibson W. (1993) *Virtual Light*, Penguin Books

Goodchild B. (1990) 'Planning and the modern/postmodern debate', *Town Planning Review*, 61 (2)

Graham S. (1995) 'The City Economy', in Healey et al. (eds) *Managing Cities, The New Urban Context*, Wiley and Sons, London

Graham S. (1995b) 'Cyberspace and the City: Issues for Planners', *Town and Country Planning*, 64(8), 198-201

Graham S (2004) Editor's introduction to 'Life After Cyberspace', in Graham S (ed) *The Cybercities Reader: Handbook for an Urban Digital Age*, Routledge, London

Graham S. and Aurigi A. (1997) 'Virtual Cities, Social Polarisation, and the Crisis in Urban Public Space', *Journal of Urban Technology*, vol.4 n.1

Graham S. and Marvin S. (1996) *Telecommunications and the City*, Routledge, London

Greenspan R. (2003) 'Another Banner E-Com Year Expected', at http://cyberatlas.internet.com/markets/retailing/article/0,,6061_2196491,00.html#table

Guidi L. (2000) 'Contributo di Leda Guidi, responsabile Iperbole', http://www.comune.bologna.it/iperbole_comunita.htm

Gumpert G. and Drucker S. (2003) 'The Perfections of Sustainability and Imperfections in the Digital Community: Paradoxes of Connection and Disconnection', paper presented at the Digital Cities 3 Workshop, Amsterdam, Sept.2003, mimeo

Guthrie K. and Dutton W. (1992) 'The Politics of Citizen Access Technology: The Development of Public Information Utilities in Four Cities', *Policy Studies Journal*, vol.20, n.4

Hanna L. (1996) 'Robin Hood turns to virtual enterprise', *The Guardian On-line* 4-1-96

Hayles K. (1993) 'The seductions of Cyberspace', in Conley V. (ed) *Rethinking Technologies*, University of Minnesota Press

Haythornthwaite and Wellman (2002) 'The Internet in Everyday Life: An Introduction', in Wellman and Haythornthwaite (eds) *The Internet in Everyday Life*, Blackwell, Malden, MA

Healey et al. (eds) (1995) *Managing Cities, The New Urban Context*, Wiley and Sons, London

Hill D.M. (1994) *Citizens and Cities*, Harvester Wheatsheaf, Hemel Hempstead

Horan T.A. (2000) *Digital Places: Building Our City of Bits*, The Urban Land Institute, Washington D.C.

Il Resto del Carlino (1995) 'C'è il postino elettronico', *Il Resto del Carlino* 16/2/95

Imbeni R. (1995) 'Che futuro sarà con l'Iperbole', *La Repubblica* 15/2/95

ISPO – Information Society Project Office (1995) 'Introduction to the Information Society: The European Way'

Jacques M (1997) 'Where the future is outrageous', *The Guardian*, 22-3-97

Jencks C. (1986) *What is Post-Modernism?*, Academy Editions, Chichester

Koolhas R and Mau B (1995) *Small, Medium, Large, Extra-Large: Office for Metropolitan Architecture*, New York: Monacelli Press

La Repubblica (1994) 'Sindaci affascinati dall'"Iperbole"', *La Repubblica* 25/6/94

La Repubblica (1996) 'E Iperbole ora naviga in tutto il mondo di Internet', *La Repubblica* 11/4/96

La Stampa (1997) 'Niente quorum, nulla di fatto per le farmacie', *La Stampa* 3/2/97

L'Unità (1996) 'Internet gratis per i cittadini: il TAR ha dato ragione al comune', *L'Unità* 27/6/96

Le Corbusier (1929) *The City of Tomorrow*, The Architectural Press, London (edition 1971)

Leary T. (1994) *Chaos and Cyberculture*, Ronin Publishing, Berkeley

Lo Piccolo F. (1995) 'Identità, permanenza, trasformazioni delle città: un'introduzione', in Lo Piccolo F. (ed) *Identità Urbana* , Gangemi Editore

Lovink G. (2004) 'The Rise and the Fall of the Digital City Metaphor and Community in 1990s Amsterdam' in Graham S (ed) *The Cybercities Reader: Handbook for an Urban Digital Age*, Routledge, London

Madani-Pour A. (1995) 'Reading the City', in Healey et al. (eds) *Managing Cities, The New Urban Context*, Wiley and Sons, London

Magnaghi A. (1995) 'L'importanza dei luoghi nell'epoca della loro dissoluzione', in Berardi F. (ed) *Cibernauti, Posturbania: la città virtuale*, Castelvecchi Editore

Maioli C. (1995) 'Comunicazione tramite elaboratore e reti civiche per i cittadini', University of Bologna, Maths Department, http://www.cs.unibo.it/~statti/einn/retimaio.html

Manchester Wireless (2003), http://www.manchesterwireless.net/

Marcoaldi F. (1993) 'La sovranita' elettronica', *La Repubblica* 3/11/93 p.31

Massey D. (1991), 'A Global Sense of Place', *Marxism Today*, June

Matteuzzi M. (1996) 'Dal comune al cittadino', *Online Magazine* Sept-Oct 96, pp.14-17

May T. (1993) *Social Research*, Open University Press, Buckingham

McBeath G. and Webb S. (1995) 'Cities, Subjectivity and Cyberspace', paper delivered to the British Sociological Association Annual Conference

Miccoli M. (1994) 'Gladiatori con la rete', *La Repubblica* 26/9/94

Mingione E. (1995) 'Social and Employment Change in the Urban Arena', in Healey et al. (eds) *Managing Cities, The New Urban Context*, Wiley and Sons, London

Mitchell D. (1995) 'The End of Public Space? People's Park, Definitions of the Public, and Democracy', Annals of the Association of American Geographers 85

Mitchell W.J. (1995) *City of Bits: Space, Place and the Infobahn*, MIT Press, Cambridge, MA

Mitchell W.J. (1999) *E-topia: 'Urban life, Jim, but not as we know it'*, MIT Press, Cambridge, MA

Mitchell W.J. (2003) ME++: The Cyborg Self and the Networked City, MIT Press, Cambridge, MA

Negroponte N. (1995) *Being Digital*, Coronet

Norberg-Schulz C. (1971) *Existence, Space and Architecture*, Studio Vista, London

Novak M. (1991) 'Liquid Architectures in Cyberspace', in Benedikt M. (ed) *Cyberspace First Steps*, The MIT Press

Ogden M.R. (1994) 'Politics in a parallel universe. Is there a future for cyberdemocracy?', *Futures* n.26

Onions I. (1997) 'Surfers ready to head for Bristol', *Evening Post*, 19/3/97

Pinch T.F. and Bijker W.E. (1987) 'The Social Construction of Facts and Artifacts: Or How the Sociology of Science and the Sociology of Technology Might Benefit Each Other', in Bijker W.E., Hughes T.P. and Pinch T.F. (eds) *The Social Construction of Technological Systems*, MIT Press, Cambridge, MA

Rheingold H. (1994) *The Virtual Community*, Secker & Warburg

Russo C. (1994) 'Bologna dialoga al computer', *Italia Oggi* 20/7/94

Sarti M. (1995) 'Internet, che Iperbole', *La Repubblica* 16/2/95

Sarti M. (1996) 'Par condicio nell'urna virtuale', *L'Unità* 28/2/96

Schiller D. (1999) *Digital Capitalism: Networking the Global Market System*, The MIT Press, Cambridge, MA

Schwartz Cowan R. (1985) 'How the refrigerator got its hum', in MacKenzie D. and Wajcman J. (eds) *The Social Shaping of Technology*, Open University Press, Bristol, PA

Shapiro A.L. (1995) 'Street Corners in Cyberspace', *The Nation* 3-7-1995

Smargiassi M. (1994) 'Il PDS lancia la cyber-politica', *La Repubblica* 16/9/94

Sorkin M. (1992) *Variations on a theme park: the new American city and the end of public space*, The Noonday Press, New York

Stallabrass J. (1995) 'Empowering Technology: The Exploration of Cyberspace', *New Left Review* 211

Stone A.R. (1991) 'Will the Real Body Please Stand Up?', in Benedikt M. (ed) *Cyberspace First Steps*, The MIT Press, Cambridge, MA

Swyngedouw E. (1993) 'Communication, mobility and the struggle for power over space', in Gianopoulos G. and.Gillespie A. (eds) *Transport and Communication Innovation in Europe*, Belhaven, London

Tatsuno S.M. (1994) 'The Multimedia City of the Future', mimeo

Toffler A. (1981) *The Third Wave*, Pan Books: London

Varesi V. (1993) 'City card, e decideranno i cittadini', *La Repubblica* 13/12/93

Virilio P. (1993) 'The Third Interval: A Critical Transition', in Conley V. (ed) *Rethinking Technologies*, University of Minnesota Press

Virilio P. (1995) 'Red alert in cyberspace!', *Radical Philosophy* 74

Visani C. (1993) 'City card in arrivo', *L'Unita'* 9/12/93

Visani C. (1994) 'Con Iperbole arriva la tele-democrazia', *L'Unità* 25/6/94

Wakeford N. (1996) 'Developing community intranets: key social issues and solutions', mimeo

Westerfield S. (1997) Digital Bristol Interactive – Final Project Report, HP Labs, Bristol

Wignall P. (1997) letter to Linda Skinner, Mimeo

Wilson E. (1995) 'The Rhetoric of Urban Space', *New Left Review* 209

Winner L. (1985) 'Do artifacts have politics?', in MacKenzie D. and Wajcman J. (eds) *The Social Shaping of Technology*, Open University Press, Bristol, PA

Yin R.K. (1994) *Case Study Research, Design and Methods*, Sage

Index